ELECTRICITY AND CONTROLS FOR HEATING, VENTILATING, AND AIR CONDITIONING

Second Edition

Ohm's Law Formulas

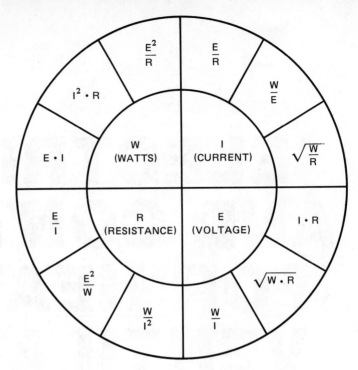

Resistor Color Code

BLACK	0
BROWN	1
RED	2
ORANGE	3
YELLOW	4
GREEN	5
BLUE	6
VIOLET	7
GRAY	8
WHITE	9

TOLERANCE

GOLD	5%
SILVER	10%
NONE	20%

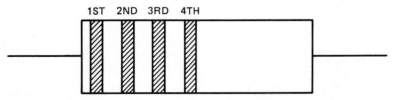

1ST   2ND   3RD   4TH

1ST BAND—NUMBER
2ND BAND—NUMBER
3RD BAND—MULTIPLIER (ADD THAT NUMBER OF 0'S TO THE FIRST TWO NUMBERS)
4TH BAND—TOLERANCE

EXAMPLE: A RESISTOR HAS COLORS OF YELLOW, VIOLET, ORANGE, AND GOLD.
FIRST TWO COLORS ARE YELLOW AND VIOLET, WHICH ARE 4 AND 7. THE THIRD BAND IS ORANGE,
WHICH IS 3. THEREFORE, ADD 3 ZEROS TO THE 47. RESISTANCE IS: 47000 OHMS.
4TH BAND IS GOLD WHICH IS 5% TOLERANCE. THE RESISTOR IS 47000 OHMS ±5%.

ELECTRICITY AND CONTROLS FOR HEATING, VENTILATING, AND AIR CONDITIONING

Second Edition

Stephen L. Herman
Bennie L. Sparkman

DELMAR PUBLISHERS INC.®

NOTICE TO THE READER

Cover design: Mary Beth Vought
Photography: Schuyler Photography
Thermostat courtesy of Honeywell Inc.
Circuit boards courtesy of Lennox/amc

Delmar Staff:
Senior Administrative Editor: David Anthony
Project Editor: Carol Micheli
Production Supervisor: Larry Main
Design Supervisor: Susan C. Mathews

For information address Delmar Publishers Inc.
2 Computer Drive West, Box 15-015,
Albany, New York 12212

Printed in the United States of America
Published simultaneously in Canada
by Nelson Canada,
A division of the Thomson Corporation

10 9 8 7 6 5 4 3 2 1

Library of Congress Cataloging in Publication Data

Herman, Stephen L.
  Electricity and controls for heating, ventilating, and
air conditioning / Stephen L. Herman, Bennie L. Sparkman.
     p.   cm.
  Includes index.
  ISBN 0-8273-4115-6
  1. Air conditioning—Electric equipment.   2. Air
conditioning—Control.   I. Sparkman, Bennie L.
II. Title.
TK4035.A35H47   1990
697—dc20   90-38093
                              CIP

CONTENTS

v

PREFACE

Electricity and Controls for Heating, Ventilating, and Air Conditioning—Second Edition is written with the assumption that the student has no prior knowledge of electricity or control systems. Basic electrical theory is presented in a practical, straightforward manner. Mathematical explanations are used only when necessary to explain certain concepts of electricity.

The text begins with the study of **basic electrical theory** and progresses to **alternating current, series and parallel circuits, and resistive, inductive, and capacitive circuits.** The text also includes information on the different types of electrical services, both single-phase and three-phase, that the technician can expect to encounter in the field. Individual devices and components common to the **air conditioning, refrigeration, and heating** field are presented in a practical manner. Devices are explained from a standpoint of how they operate and how they are used. The text contains **testing procedures** for many of the devices covered.

Both single-phase and three-phase motors are presented in a practical "how it works" manner. The types of single-phase motors covered are **shaded-pole induction, split-phase, and multi-speed motors.** The types of three-phase motors covered are **squirrel cage induction, wound rotor induction, and synchronous.** The types of single-phase split-phase motors covered are **resistance-start induction run, capacitor-start induction run, and permanent-split capacitor motors.** A separate unit is included to discuss different methods of variable speed control for AC motors.

Control circuits are developed using the components discussed in the text. The text assumes that the student has no prior knowledge of control systems. **Operation of manufacturers' control schematics is explained to aid the student in understanding how a control system operates and how to troubleshoot the system.**

Electricity and Controls for Heating, Ventilating, and Air Conditioning—Second Edition includes a new section on **ice makers.** It discusses common types of residential and commercial ice makers and their control systems. A step-by-step description explains the operation of their circuits.

Solid-state devices common to the HVAC field are covered in a straightforward manner. The devices covered are **diodes, transistors, SCRs, diacs, triacs, and operational amplifiers.** The last section of the text covers **programmable logic controllers,** which are becoming more and more common in industry.

ACKNOWLEDGMENTS

The authors of this text would like to express appreciation to the following people for their input as reviewers of the text:

Norman Easton
Genesee Area Skill Center
Flint, Michigan

Gred Jourdan
Wenatchee Valley College
Wenatchee, Washington

James D. Bussey
Griffin Technical Institute
Griffin, Georgia

Loyd Cossey
Ozarka Vocational Technical School
Melbourne, Arizona

SECTION 1
Basic Electricity

UNIT 1
Atomic Structure

To understand electricity, it is necessary to start with the study of atoms. The atom is the basic building block of the universe. All materials are made from a combination of atoms. An atom is the smallest part of an element. The three principal parts of an atom are the *electron*, the *neutron*, and the *proton*. Figure 1-1 illustrates these parts of the atom. Notice that the proton has a positive charge, the electron has a negative charge, and the neutron has no charge. The neutron and proton combine to form the *nucleus* of the atom. The electron orbits around the outside of the nucleus. Notice that the electron is shown to be larger than the proton. Actually, the electron is about three times larger than a proton, but the proton weighs about 1840 times more than an electron. It is like comparing a soap bubble to a piece of buckshot. This means that the proton is a very massive particle in comparison to the electron.

To understand atoms, it is necessary to first understand two basic laws of physics. One of these laws is the law of charges that states that opposite charges attract and like charges repel. Figure 1-2 illustrates this principle. In figure 1-2, charged balls are suspended from strings. Notice that the two balls that contain opposite charges are attracted to each other. The two positively charged balls and the two negatively charged balls are repelled from each other. Since the proton has a positive charge and the electron has a negative charge, they are attracted to each other.

The second law that must be understood is the law of centrifugal force. This law states that a spinning object will pull away from its centerpoint. The faster an object spins, the greater the centrifugal force becomes. Figure 1-3 shows an example of this principle. If an object is tied to a string, and the object is spun around, it will try to pull away from you. The faster the object spins, the greater the force is that tries to pull the object away. Cen-

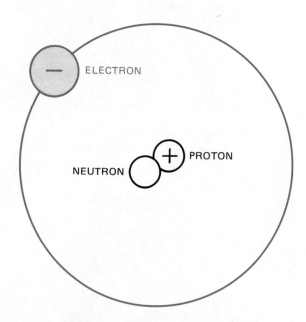

FIGURE 1-1 Principal parts of an atom

2

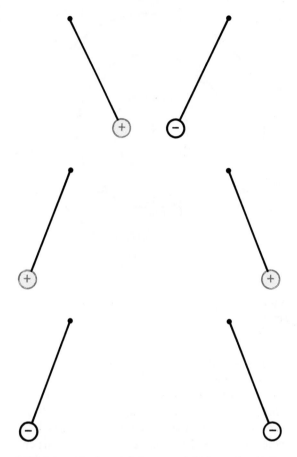

FIGURE 1-2 The law of charges states that opposite charges attract and like charges repel.

trifugal force keeps the electron from falling into the nucleus of the atom. The faster an electron spins, the farther away from the nucleus it will be.

Atoms have a set number of electrons that can be contained in one orbit or shell. The number of electrons that can be contained in any one shell is found by the formula $(2N^2)$. The letter "N" represents the number of the orbit or shell. For example, the first orbit can hold no more than two electrons. $2 \times (1)^2 = 2 \times 1 = 2$. The second orbit can hold no more than 8 electrons. $2 \times (2)^2 = 2 \times 4 = 8$. The third orbit can contain not more than 18 electrons. $2 \times (3)^2 = 2 \times 9 = 18$. The fourth orbit cannot hold more than 32 electrons. $2 \times (4)^2 = 2 \times 16 = 32$.

The outer shell of an atom is known as the *valence shell*. Any electrons located in the outer shell of an atom are known as *valence electrons*. The valence shell of an atom cannot hold more than eight electrons. It is the valence electrons that are of primary concern in the study of electricity, because it is these electrons that explain much of electrical theory. A conductor, for instance, is made from a material that contains one or two valence electrons. When an atom has only one or two valence electrons, they are loosely held by the atom and are easily given up for current flow. Silver, copper, and aluminum all contain one valence electron. Although all of these materials contain only one valence electron, silver is a better conductor than copper, and copper is a better conductor than aluminum. The reason for this is that an atom of silver is larger than an atom of copper, and an atom of copper is larger than an atom of aluminum. Since an atom of silver is larger than an atom of copper, it contains more orbits than an atom of copper. This means that the valence electron of silver is farther away from the nucleus than an atom of copper. Since the speed an electron spins is determined by its distance from the nucleus, the va-

FIGURE 1-3 Centrifugal force causes an object to pull away.

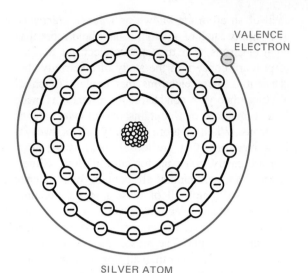

VALENCE
ELECTRON

SILVER ATOM

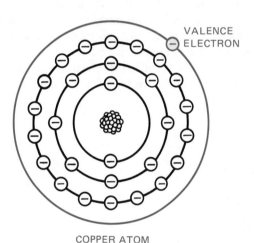

VALENCE
ELECTRON

COPPER ATOM

FIGURE 1-4

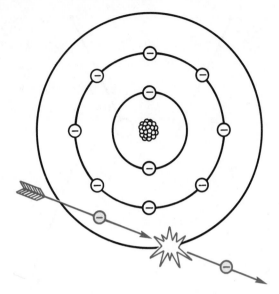

FIGURE 1-5 An electron knocked out of orbit by another electron

Electricity is the flow of electrons. It is produced by knocking the electrons of an atom out of orbit by another electron. Figure 1-5 illustrates this action. When an atom contains only one valence electron, it is easily given up when it is struck by another electron. The striking electron gives its energy to the electron being struck. The striking electron settles into orbit around the atom, and the

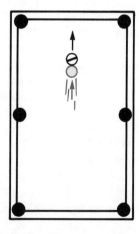

FIGURE 1-6 The cue ball gives energy to the stationary ball.

lence electron of silver is spinning around the nucleus at a faster speed than the valence electron of copper. Therefore, the valence electron of silver contains more energy than the valence electron of copper. When the valence electron of silver is knocked out of orbit, it simply contains more energy than the valence electron of copper, and therefore, makes a better conductor of electricity. Copper is a better conductor of electricity than aluminum for the same reason. Figure 1-4 shows an atom of silver and an atom of copper.

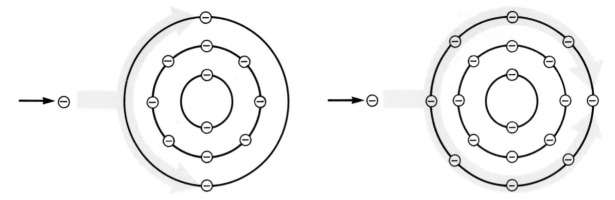

FIGURE 1-7 Energy is divided between two valence electrons.

FIGURE 1-8 Energy is divided among eight electrons.

electron that was struck moves off to strike another electron. This same action can often be seen in the game of pool. If the moving cue ball strikes a stationary ball exactly right, the energy of the cue ball is given to the stationary ball. The stationary ball then moves off with most of the energy of the cue ball, and the cue ball stops moving. Figure 1-6 illustrates this condition. Notice that the stationary ball did not move off with the same energy of the cue ball. It moved off with most of the energy of the cue ball. Some of the energy of the cue ball was lost to heat when it struck the stationary ball. This is true when one electron strikes another also. This is the reason that a wire heats when current flows through it. If too much current flows through a wire, it will overheat and damage the wire or become a fire hazard.

If an atom that contains two valence electrons is struck by a moving electron, the energy of the striking electron is divided between the two valence electrons. Figure 1-7 shows this action. If the valence electrons are knocked out of orbit, they will contain only half the energy of the striking electron. This action can also be seen in the game of pool. If a moving cue ball strikes two stationary balls at the same time, the energy of the cue ball is divided between the two stationary balls. Both of the stationary balls will move, but with only half the energy of the cue ball.

Materials that are made from atoms that contain seven or eight valence electrons are known as insulators. *Insulators* are materials that resist the flow of electricity. Some good examples of insulator materials are rubber, plastic, glass, and wood. Figure 1-8 illustrates what happens when a moving electron strikes an atom that contains eight valence electrons. The energy of the moving electron is divided so many times that it has little effect on the atom.

REVIEW QUESTIONS

1. What are the three subatomic parts of atoms and what charge does each carry?

2. How many times larger is an electron than a proton?

3. The weight of a proton is how many times heavier than that of an electron?

4. State the law of charges.

5. What force keeps the electron from falling into the nucleus of the atom?

6. Materials that make the best conductors contain how many valence electrons?

7. Materials that make the best insulators contain how many valence electrons?

8. What is electricity?

UNIT 2

Electrical Quantities and Ohm's Law

Electricity has a standard set of values. Before a person can work with electricity, that person must have a knowledge of these values and how to use them. Since the values of electrical measurement have been standardized, they are understood by everyone who uses them. For instance, carpenters use a standard system for measuring length, such as the inch or foot. Imagine what a house would look like that was constructed by two carpenters who used different lengths of measure for an inch or foot. The same holds true for people who work with electricity. A volt, ampere, or ohm is the same for everyone who uses them.

COULOMB

A *coulomb* is a quantity measurement of electrons. One coulomb contains 6.25×10^{18} electrons. The number shown in figure 2-1 is the number of electrons in one coulomb. Since the coulomb is a quantity measurement, it is similar to a quart, gallon, or liter. It takes a certain amount of liquid to equal a quart, just as it takes a certain amount of electrons to equal a coulomb.

6,250,000,000,000,000,000

FIGURE 2-1

AMPERE

The *ampere*, or amp, is defined as one coulomb per second. Notice that the definition of an amp involves a quantity measurement (the coulomb) combined with a time measurement (the second). One amp of current flows through a wire when one coulomb flows past a point in one second, figure 2-2. The ampere is a measurement of the actual amount of electricity that is flowing through a circuit. In a water system, it would be comparable to gallons per minute or gallons per second, figure 2-3.

VOLT

Voltage is actually defined as *electromotive force*, or EMF. It is the force that pushes the electrons through a wire and is often referred to as electrical pressure. One should remember that voltage cannot flow. To say that voltage flows is like saying that pressure flows through a pipe. Pressure can push water through a pipe, and it is correct to say that water flows through a pipe, but it is not correct to say that pressure flows through a pipe. The same is true for voltage. The voltage pushes current through a wire, but voltage cannot flow through a wire. In a water system, the voltage could

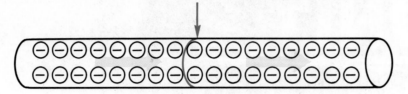

FIGURE 2-2 One coulomb flowing past a point in one second

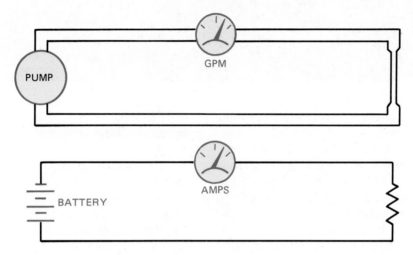

FIGURE 2-3 Water flow is similar to current flow.

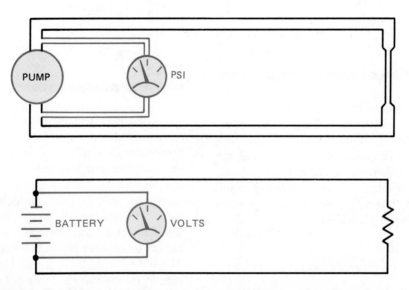

FIGURE 2-4 The pressure in a water system is comparable to the voltage in an electric circuit.

SECTION 1   BASIC ELECTRICITY

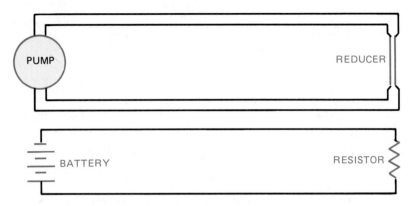

FIGURE 2-5 A reducer reduces the flow of water through a water system just as a resistor reduces the flow of current in an electric circuit.

be compared to the pressure of the system, figure 2-4.

OHM

The *ohm* is the measure of the resistance to the flow of current. The voltage of the circuit must overcome the resistance before it can cause electrons to flow through it. Without resistance, every electrical circuit would be a short circuit. All electrical loads, such as heating elements, lamps, motors, transformers, and so forth, are measured in ohms. In a water system, a reducer could be used to control the flow of water. In an electrical circuit, a resistor can be used to control the flow of electrons. Figure 2-5 illustrates this concept.

WATT

Wattage is a measure of the amount of power that is being used in the circuit. It is proportional to the amount of voltage and the amount of current flow. To understand watts, return to the example of the water system. Assume that a water pump has a pressure of 120 psi (pounds per square inch) and causes a flow rate of one gallon per second. Now assume that this water is used to drive a waterwheel as shown in figure 2-6. Notice that the waterwheel has a radius of 1 ft from the center shaft to the rim of the wheel. Since water weighs 8.34

pounds per gallon, and is being forced against the wheel at a pressure of 120 psi, the wheel could develop a torque of 1000.8 foot pounds (120 × 8.34 × 1 = 1000.8). If the pressure is increased to 240 psi, but the water flow remains constant, the force against the wheel will double (240 × 8.34 × 1 = 2001.6). If the pressure remains at 120 psi, but the water flow is increased to two gallons per second, the force against the wheel will again double (120 × 16.68 × 1 = 2001.6). Notice that the amount of power developed by the waterwheel is determined by both the amount of pressure driving the water and the amount of flow.

The power of an electrical circuit is very similar. Figure 2-7 shows a resistor connected to a circuit with a voltage of 120 volts and a current flow of 1 amp. The resistor shown represents an electric heating element. When 120 volts forces a current

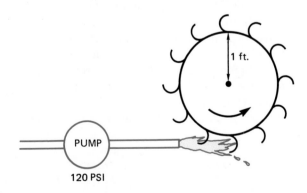

FIGURE 2-6 Pump used to drive a waterwheel

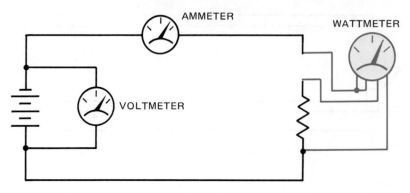

FIGURE 2-7 The amount of voltage and current determine the power

of 1 amp through it, the heating element will produce 120 watts of heat (120 × 1 = 120 watts). If the voltage is increased to 240 volts, but the current remains constant, the element will produce 240 watts of heat (240 × 1 = 240 watts). If the voltage remains at 120 volts, but the current is increased to 2 amps, the heating element will again produce 240 watts (120 × 2 = 240). Notice that the amount of power used by the heating element is determined by the amount of current flow and the voltage driving it.

OHM'S LAW

Ohm's Law is named for the German scientist, George S. Ohm. Ohm discovered that all electrical quantities are proportional to each other and can therefore be expressed as mathematical formulas. In its simplest form, Ohm's Law states that *it takes 1 volt to push 1 amp through 1 ohm.*

Figure 2-8 shows three basic Ohm's Law formulas. In these formulas, "*E*" stands for electromotive force and is used to represent the voltage. The "*I*" stands for the intensity of the current and is used to represent the amount of current flow or amps. The letter "*R*" stands for resistance and is used to represent the ohms.

The first formula states that the voltage can be

$$E = I \times R \qquad I = \frac{E}{R} \qquad R = \frac{E}{I}$$

FIGURE 2-8

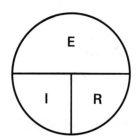

FIGURE 2-9 Ohm's Law formula chart

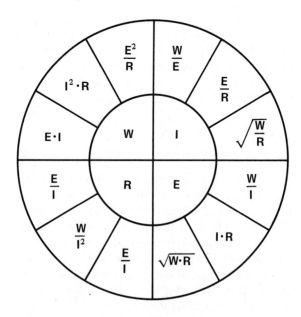

FIGURE 2-10 Formula chart used to find volts, amps, resistance, and watts

found if the current and resistance are known. Voltage is equal to amps multiplied by ohms. For example, assume a circuit has a resistance of 50 ohms and a current flow through it of 2 amps. The voltage connected to this circuit is 100 volts (2 amps × 50 ohms = 100 volts). The second formula indicates that if the voltage and resistance of the circuit are known, the amount of current flow can be found. Assume a 120-volt circuit is connected to a resistance of 30 ohms. The amount of current that will flow in the circuit is 4 amps (120 volts/30 ohms = 4 amps). The third formula states that if the voltage and current flow in a circuit are known, the resistance can be found. Assume a circuit has a voltage of 240 volts and a current flow of 10 amps. The resistance in the circuit is 24 ohms (240 volts/10 amps = 24 ohms).

Figure 2-9 shows a simple chart that can be a great help when trying to remember an Ohm's Law formula. To use the chart, cover the quantity that is to be found. For example, if the voltage, "E," is to be found, cover the "E" on the chart. The chart now shows the remaining letters "IR." $E = I \times R$. If the current is to be found, cover the "I" on the chart. The chart now shows E/R. $I = E/R$. If the resistance of a circuit is to be found, cover the "R" on the chart. The chart now shows E/I. $R = E/I$. A larger chart that shows the formulas needed to find watts as well as the voltage, amperage, and resistance is shown in figure 2-10.

REVIEW QUESTIONS

1. What is a coulomb?

2. What is the definition of an amp?

3. Define the term voltage.

4. Define the term ohm.

5. Define the term watt.

6. An electric heating element has a resistance of 16 ohms and is connected to a voltage of 120 volts. How much current (amps) will flow in the circuit?

7. How many watts of heat are being produced by the heating element in question #6?

8. A 240-volt circuit has a current flow of 20 amps. How much resistance (ohms) is connected in the circuit?

9. An electric motor has an apparent resistance of 15 ohms. If a current of 8 amps is flowing through the motor, what is the connected voltage?

10. A 240-volt air-conditioning compressor has an apparent resistance of 8 ohms. How much current (amps) will flow in the circuit?

11. How much power (watts) is being used by the compressor in question #8?

12. A 5000-watt electric heating unit is connected to a 240-volt line. What is the current flow (amps) in the circuit?

13. If the voltage in question #12 is reduced to 120 volts, how much current (amps) would be needed to produce the same amount of power?

14. Is it less expensive to operate the electric heating unit in question #12 on 240 volts or on 120 volts? Explain your answer.

UNIT 3
Measuring Instruments

Anyone desiring to work in the air-conditioning and refrigeration field must become proficient with the common instruments used to measure electrical quantities. These instruments are the voltmeter, ammeter, and ohmmeter. In the air-conditioning and refrigeration field, the technician works almost exclusively with alternating current. For this reason, the meters covered in this unit are intended to be used in an AC system.

VOLTMETER

The voltmeter is designed to be connected directly across the source of power. Figure 3-1 shows a voltmeter being used to test the voltage of a panel box. Notice that the leads of the meter are connected directly across the source of voltage. The reason a voltmeter can be connected directly across the power line is because it has a very high resistance connected in series with the meter movement, figure 3-2. A common resistance for a voltmeter is about 20,000 ohms per volt for DC and 5,000 ohms per volt AC. Assume the voltmeter shown in figure 3-2 is an AC meter and has a full-scale range of 300 volts. The resistor connected in series with the meter would, therefore, have a resistance of 1,500,000 ohms (300 volts × 5000 ohms per volt = 1,500,000 ohms).

Most voltmeters are multi-ranged, which means that they are designed to use one meter movement to measure several ranges of voltage. For example, one meter may have a selector switch that permits full-scale ranges to be selected. These ranges may be 3 volts full-scale, 12 volts full-scale, 30 volts full-scale, 60 volts full-scale, 120 volts full-scale, 300 volts full-scale, and 600 volts full-scale. The

FIGURE 3-1 Voltmeter being used to test the voltage of a panel

reason for making a meter with this number of scales is to make the meter as versatile as possible. If it is necessary to check for a voltage of 480 volts, the meter can be set on the 600-volt range. If it becomes necessary to check a control voltage of 24 volts, however, it would be very difficult to do on the 600-volt range. If the meter is set on the 30-volt range, however, it becomes a simple matter to test for a voltage of 24 volts. The meter shown in figure 3-3 has multi-range selection for voltage.

When the selector switch of this meter is turned, steps of resistance are inserted in the circuit to increase the range, or removed from the circuit to decrease the range, figure 3-4. Notice that when the higher voltage settings are selected, more resistance is inserted in the circuit.

Another type of voltmeter that is gaining popularity is the digital meter. A digital meter displays the voltage in digits instead of using a meter movement, figure 3-5. Digital meters have several advantages over voltmeters that use a meter movement (commonly called analog meters). The greatest advantage is that the input impedance, or resistance, is higher. Analog meters commonly have a resistance of about 5,000 ohms per volt. This means

that on a 3-volt full-scale range, the meter movement has a resistance of 15,000 ohms connected in series with it ($3 \times 5,000 = 15,000$). On the 600-volt full-scale range, the meter movement has a resistance of 3,000,000 ohms connected in series with it ($600 \times 5,000 = 3,000,000$). Digital meters commonly have an input impedance of 10,000,000 ohms (10 megohms), regardless of the range they are set on. The advantage of this high input impedance is that it does not interfere with a low-power circuit. The advantage of this may not be too clear at first, because most technicians are used to working with circuits that have more than enough power to operate the meter. However, many of the newer controls are electronic; these circuits may be greatly altered if tested with a low impedance meter.

For example, assume an electronic control is operated on 5 volts, and has a total current capacity of 100 microamps (.000100 amps). Now assume that a 5,000-ohm-per-volt meter is set on the 12-volt scale and is to be used to check the control circuit. The voltmeter has a resistance of $5,000 \times 12 = 60,000$ ohms. If this meter is used to test the circuit, the meter will have a current draw of 83.3 microamps (5 volts/60,000 ohms = .0000833

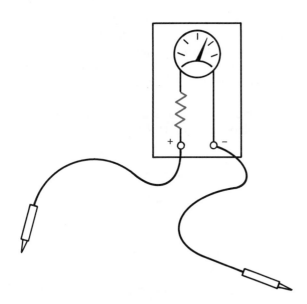

FIGURE 3-2 A voltmeter has high resistance connected in a series with the meter movement.

FIGURE 3-3 Multi-meter with an analog scale (Courtesy of Triplett Corp.)

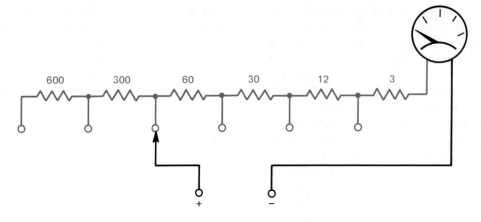

FIGURE 3-4   A multi-range voltmeter

amps). Since the control circuit only has a total current capacity of 100 microamps, the meter is using most of the current to operate. The circuit has been changed to such a degree that it can no longer operate.

If the digital meter is used to test this same circuit, it will have a current draw of .5 microamp (5 volts/10,000,000 = .00000005 amps). The circuit will be able to furnish the .5 microamp needed to operate the meter without a problem or altering the circuit.

Another advantage of the digital meter is that it is generally easier for an inexperienced person to learn to read. Analog meters can be used for about 99% of the measurements that must be taken, but it generally takes some time and practice to read them properly.

Learning to read the scale of a multi-meter takes time and practice. Most people use meters every day without thinking about it. A very common type of meter used daily by most people is shown in figure 3-6. The meter illustrated is a speedometer similar to those seen in an automobile. This meter is designed to measure speed. It is calibrated in miles per hour (mph). The speedometer shown has a full-scale value of 80 mph. If the pointer is positioned as shown in figure 3-6, most people would know instantly that the speed of the automobile is 55 mph.

Figure 3-7 illustrates another common meter used by most people. This meter is used to mea-

FIGURE 3-5   Digital Multi-meter (Courtesy of Triplett Corp.)

FIGURE 3-6   Speedometer

FIGURE 3-7 Fuel Gauge

sure the amount of fuel in the tank of an automobile. Most people can glance at the pointer of the meter and know that the meter is indicating that there is 1/4 of a tank of fuel remaining. Now assume that the tank has a capacity of 20 gallons. The meter is now indicating that there are a total of 5 gallons of fuel remaining in the tank.

Learning to read the scale of a multi-meter is similar to learning to read a speedometer or fuel gauge. The meter scale shown in figure 3-8 has several scales used to measure different values and quantities. The very top of the scale is used to measure resistance or ohms. Notice that the scale begins at the left-hand side with infinity, and zero can be found at the far right-hand side. Ohmmeters will be covered later in this unit. The second scale is labeled AC—DC and is used to measure voltage. Notice this scale has three different full-scale values. The top scale is 0–300, the second scale is 0–60, and the third scale is 0–12. The scale used is determined by the setting of the range control switch. The third set of scales is labeled AC AMPS.

This scale is used with a clamp-on ammeter attachment that can be used with some meters. The last scale is labeled dbm and is seldom if ever used by the technician in the field.

Reading a Voltmeter

Notice that the three voltmeter scales use the primary numbers 3, 6, and 12 and are in multiples of 10 of these numbers. Since these numbers are in multiples of 10, it is an easy matter to multiply or divide the readings in your head by moving a decimal point. Remember that any number can be multiplied by 10 by moving the decimal point one place to the right, and any number can be divided by 10 by moving the decimal point one place to the left. For example, if the selector switch is set to permit the meter to indicate a voltage of 3 volts full-scale, the 300-volt scale would be used, and the reading divided by 100. The reading can be divided by 100 by moving the decimal point 2 places to the left. In figure 3-9, the meter is indicating a voltage of 2.5 volts if the selector switch is set for 3 volts full-scale. The pointer is indicating a value of 250. Moving the decimal point 2 places to the left will give a reading of 2.5 volts. If the selector switch is set for a full-scale value of 30 volts, the meter shown in figure 3-9 would be indicating a value of 25 volts. This reading is obtained by dividing the scale by 10 and moving the decimal point one place to the left.

Now assume that the meter has been set to have a full-scale value of 600 volts. The meter

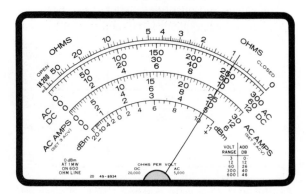

FIGURE 3-8 Multi-meter face

FIGURE 3-9

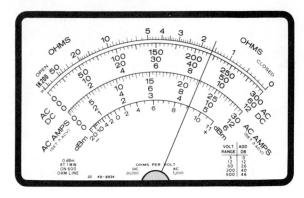

FIGURE 3-10

shown in figure 3-10 is indicating a voltage of 440 volts. Since the full-scale value of the meter is set for 600 volts, use the 60-volt range and multiply the reading on the meter by 10. This can be done by moving the decimal point one place to the right. The pointer in figure 3-10 is indicating a value of 44. If this value is multiplied by 10, the correct voltage reading becomes 440 volts.

There are three distinct steps that should be followed when reading a meter. This is especially true for someone who has not had a great deal of experience reading a multi-meter. These steps are:

1. DETERMINE WHAT THE METER INDICATES. Is the meter set to read a value of DC voltage, DC current, AC voltage, AC current, ohms? It is impossible to read a meter if you do not know what the meter is measuring.

2. DETERMINE THE FULL-SCALE VALUE OF THE METER. The advantage of a multi-meter is that it has the ability to measure a wide range of values and quantities. After it has been determined what quantity the meter is set to mea-

sure, it must then be determined what the range of the meter is. There is a great deal of difference in readings when the meter is set to indicate a value of 600 volts full-scale and when it is set for 30 volts full-scale.

3. READ THE METER. The last step is to determine what the meter is indicating. It may be necessary to determine the value of the hatch marks on the meter face for the range the selector switch is set for. If the meter in figure 3-8 is set for a value of 300 volts full-scale, each hatch mark has a value of 5 volts. If the full-scale value of the meter is 60 volts, however, each hatch mark has a value of 1 volt.

AMMETER

The ammeter, unlike the voltmeter, is a very low impedance device. The ammeter must be connected in series with the load to permit the load to limit the current flow, figure 3-11. An ammeter has a typical impedance of less than .1 ohm. If this meter is connected in parallel with the power supply, the impedance of the ammeter is the only thing to limit the amount of current flow in the circuit. Assume that an ammeter with an impedance of .1 ohm is connected across a 240-volt AC line. The current flow in this circuit would be 2400 amps (240/.1 = 2400). The blinding flash of light would

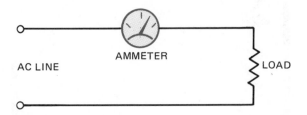

FIGURE 3-11 An ammeter must be connected in series with the load.

FIGURE 3-12 AC inline ammeter

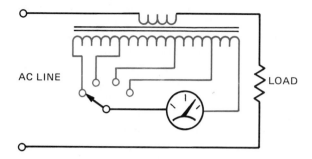

FIGURE 3-13 A current transformer provides different scales.

be followed by the destruction of the ammeter. Ammeters that are connected directly into the circuit as shown in figure 3-11 are referred to as *in-line* ammeters. Figure 3-12 shows an inline ammeter.

Notice that the meter in figure 3-12 has several ranges. AC ammeters use a current transformer to provide multi-scale capability. The primary of the transformer is connected in series with the load, and the ammeter is connected to the secondary of the transformer. Figure 3-13 illustrates this type of connection. Notice that the range of the meter is changed by selecting different taps on the secondary of the current transformer. The different taps on the transformer provide different turns ratios between the primary and secondary of the transformer.

When a large amount of AC current must be measured, a current transformer is connected into the power line. The ammeter is then connected to the secondary of the transformer. The AC ammeters are designed to indicate a current of 5 amps, and the current transformer determines the value of line current that must flow to produce a current of 5 amps on the secondary of the transformer. The incoming line may be looped around the opening in the transformer several times to produce the proper turns ratio between the primary and the secondary windings. Figure 3-14 shows a transformer of this type. This type of connection is often used for panel meters mounted on large commercial units.

The type of ammeter used in the field by most air-conditioning service technicians is the clamp-on type of ammeter similar to the one shown in figure 3-15. To use this meter, the jaw of the meter is clamped around one of the conductors supplying power to the load. Figure 3-16 shows this connection. Notice that the meter is clamped around only one of the power lines. If the meter is clamped around more than one line, the magnetic fields of the wires cancel each other and the meter indicates zero.

This type of meter also uses a current trans-

FIGURE 3-14 Current transformer used to meter large AC currents

FIGURE 3-15 Multi-meter and clamp-on ammeter combination (Courtesy of Triplett Corp.)

UNIT 3 MEASURING INSTRUMENTS

FIGURE 3-16 Clamp-on ammeter being used to measure the running current of a compressor

former to operate the meter. The jaw of the meter is part of the core material of the transformer. When the meter is connected around the current-carrying wire, the changing magnetic field produced by the AC current induces a voltage into the current transformer. The strength of the magnetic field and its frequency determines the amount of voltage induced in the current transformer. Since 60 Hz is a standard frequency throughout the country, the amount of induced voltage is proportional to the strength of the magnetic field.

The clamp-on type of ammeter can have different range settings by changing the turns ratio of the secondary of the transformer just as the inline ammeter does in figure 3-13. The primary of the current transformer is the conductor the ammeter is connected around. If the ammeter is connected around one wire as shown in figure 3-16, the primary has one turn of wire as compared to the number of turns of wire in the secondary. If two turns of wire are wrapped around the jaw of the am-

meter, the primary winding now contains two turns instead of one, and the turns ratio of the transformer is changed, figure 3-17. The ammeter will now indicate double the amount of current in the circuit. The reading on the scale of the meter would have to be divided by two to get the correct reading. For example, assume two turns of wire have been wrapped around the ammeter jaw, and the meter indicates a current of 3 amps. The actual current in this circuit is 1.5 amps ($3 \div 2 = 1.5$). The ability to change the turns ratio of a clamp-on ammeter can be very useful for measuring low currents such as those found in a control circuit. Changing the turns ratio of the transformer is not limited to wrapping two turns of wire around the jaw of the ammeter. Any number of turns of wire can be wrapped around the jaw of the ammeter and the reading will be divided by that number. If three turns of wire are wrapped around the jaw of the meter, the reading will be divided by three.

A very handy device can be made by wrapping 10 turns of wire around some core of nonmagnetic material, such as a thin piece of plastic pipe. Since the device is intended to be used for low current readings, the wire size does not have to be large. A #20 or #18 American Wire Gauge (AWG) wire is large enough. Plastic tape is used to secure the turns of wire to the core material, and two alligator clips are connected to the ends of the wire. This device is shown in figure 3-18. To use this device, break connection in the circuit to be tested, and insert the 10 turns of wire using the alligator clips. The jaw of the ammeter is placed around the plastic core. The primary of the trans-

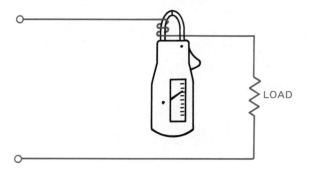

FIGURE 3-17 Two turns of wire change the turns ratio of the transformer

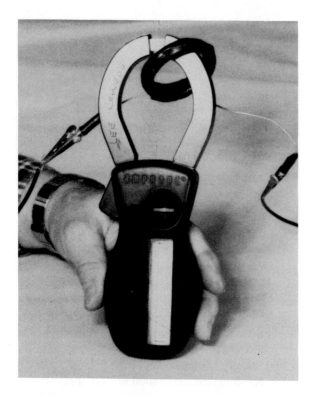

FIGURE 3-18 Scale divider used with clamp-on ammeter

measure AC volts and ohms as well as AC current. This makes the meter a more versatile instrument. When voltage or resistance is to be measured, a set of leads is attached to the meter.

OHMMETER

The ohmmeter is used to measure resistance. The common volt-ohm-milliammeter (VOM) contains an ohmmeter. The ohmmeter must provide its own power supply to measure resistance. This is done with batteries located inside the instrument. When resistance is to be measured, the meter must first be zeroed. This is done with the ohms adjust control located on the front of the meter. To zero the meter, connect the leads together and adjust the ohms adjust knob until the meter indicates 0 at the far right end of the scale, figure 3-20. When the leads are separated, the meter will again indicate infinity resistance at the far left side of the meter

former now contains 10 turns of wire, and the scale factor can now be divided by a factor of 10. The correct ammeter reading is found by moving the decimal point one place to the left. If the ammeter has a low scale of 6 amps full-scale, it can now be used to measure .6 amps full-scale. This can be a real advantage when it is necessary to measure control currents that may not be greater than .2 amps under normal operating conditions.

Some clamp-on ammeters use a digital readout instead of a meter movement. A digital type meter is shown in figure 3-19. The digital ammeters are generally better for measuring low current values, but the 10 turn scale divider can be used with these ammeters also. Just remember to divide the reading shown by a factor of 10. The clamp-on ammeters discussed in this unit are intended to be used for measuring AC currents only, and will not operate if connected to a DC line. There are clamp-on type ammeters, however, that can be used to measure DC current.

Many clamp-on ammeters are designed to

FIGURE 3-19 Digital Clamp-on Ammeter (Courtesy of Simpson Electric Co.)

FIGURE 3-20 The ohmmeter must be set at zero.

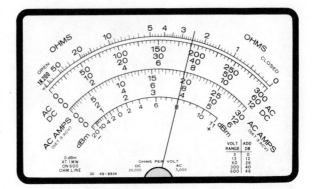

FIGURE 3-21

scale. When the leads are connected across a resistance, the meter will again indicate up scale. Figure 3-21 shows a meter indicating a resistance of 25 ohms, assuming the rang setting is R × 10.

Ohmmeters can have different range settings such as R × 1, R × 100, R × 1000, or R × 10,000. On the R × 1 setting, the resistance is measured straight off the resistance scale located at the top of the meter. If the range is set for R × 1000, however, the reading must be multiplied by 1000. The ohmmeter reading shown in figure 3-21 would be indicating a resistance of 25,000 ohms if the range had been set for R × 1000. Notice that the ohmmeter scale is read backward from the other scales. Zero ohms is located on the far right side of the scale, and maximum ohms is located at the far left side. It generally takes a little time and practice to read the ohmmeter properly.

Digital ohmmeters display the resistance in figures instead of using a meter movement. When using a digital ohmmeter, care must be taken to notice the scale indication on the meter. For example, most digital meters will display a "K" on the scale to indicate kilohms or an "M" to indicate megohms. (*Kilo* means 1000, and *mega* means 1,000,000.) If the meter is showing a resistance of (.200 K), it means .2 × 1000 or 200 ohms. If the meter indicates (1.65 M), it means 1.65 × 1,000,000 or 1,650,000 ohms.

The ohmmeter must never be connected to a circuit with power on. Since the ohmmeter uses its own internal power supply, it has a very low operating voltage. If a meter is connected to power when it is set in the ohms position, it will probably damage or destroy the meter.

REVIEW QUESTIONS

1. What type of meter has a high resistance connected in series with the meter movement?

2. How is a voltmeter connected into the circuit?

3. If a voltmeter has a resistance of 5000 ohms per volt, what is the resistance of the meter when it is set on the 300-volt range?

4. What is the advantage of using a voltmeter that has a high impedance as opposed to a low-impedance meter?

5. What is an analog meter?

6. Why must an ammeter be connected in series with the load?

SECTION 1   BASIC ELECTRICITY

7. What device is used to change the scale values of an AC ammeter?

8. What is meant by the term "inline" ammeter?

9. A clamp-on ammeter has three turns of wire wrapped around the movable jaw. If the meter is indicating a current of 15 amps, how much current is actually flowing in the circuit?

10. List the three steps for reading a meter.

11. What type of meter contains its own internal power supply?

12. What precaution must be taken when using an ohmmeter?

UNIT 4

Electrical Circuits

Electrical circuits can be divided into three basic types. These are series, parallel, and combination. The simplest of these circuits is the series circuit shown in figure 4-1. A series circuit is characterized by the fact that it has only one path for current flow. If it is assumed that current must flow from point "A" to point "B" in the circuit shown in figure 4-1, it will flow through each of the resistors. Therefore, *the current flow in a series circuit must be the same at any point in the circuit.* Another rule of series circuits states that *the sum of the voltage drops around the circuit must equal the applied voltage.* A third rule of series circuits states that *the total resistance is equal to the sum of the individual resistors.*

The circuit shown in figure 4-2 shows the values of current flow, voltage drop, and resistance for each of the resistors. Notice that the total resistance of the circuit can be found by adding the

values of each of the individual resistors (20 + 10 + 30 = 60 ohms). The amount of current flow in the circuit can be found by using Ohm's Law.

$$I = \frac{E}{R}$$
$$I = \frac{120}{60}$$
$$I = 2 \text{ amps}$$

There is a current flow in the circuit of 2 amps. Notice that the same current flows through each of the resistors. The voltage drop across each resistor can be found using Ohm's Law ($E = I \times R$). The voltage dropped across resistor R1 is $2 \times 20 = 40$ volts. This means that it takes a voltage of 40 volts to push 2 amps of current through 20 ohms of resistance. If a voltmeter is connected across resistor R1, it would indicate a voltage drop of 40 volts.

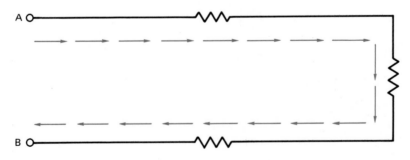

FIGURE 4-1 Series circuit

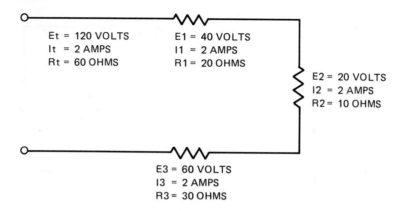

Et = 120 VOLTS
It = 2 AMPS
Rt = 60 OHMS

E1 = 40 VOLTS
I1 = 2 AMPS
R1 = 20 OHMS

E2 = 20 VOLTS
I2 = 2 AMPS
R2 = 10 OHMS

E3 = 60 VOLTS
I3 = 2 AMPS
R3 = 30 OHMS

FIGURE 4-2

The voltage drop of resistor R2 can be found the same way ($E = I \times R$), ($2 \times 10 = 20$). The voltage dropped across resistor R2 is 20 volts. The third resistor has a voltage drop of $2 \times 30 = 60$ volts. Notice that if the voltage drops are added together, they will equal the voltage applied to the circuit ($40 + 20 + 60 = 120$).

Since a series circuit has only one path for current flow, if any point in the circuit should become open, current flow throughout the entire circuit will stop. Some strings of christmas tree lights are wired in series. If any bulb in the string burns out, all of the lights will go out. When the defective bulb is replaced, all of the lights will operate. Because of this characteristic of series circuits, fuses and circuit breakers are connected in series with what they are intended to protect. Figure 4-3 shows a fuse used to protect an air-conditioning unit. If the fuse should open, current flow to the entire circuit will stop.

PARALLEL CIRCUITS

Parallel circuits are characterized by the fact that they have more than one path for current flow. The circuit shown in figure 4-4 illustrates this. If the current in this circuit is assumed to flow from point "A" to point "B," there are three separate paths through which it can flow. Current can flow from point "A," through resistor R1 to point "B," or it can flow from point "A" through resistor R2 to point "B," or from point "A" through resistor

R3 to point "B." Since current can flow through each of these resistors, the total current flow in the circuit is the sum of these individual currents. A rule for parallel circuits states that *the total current in a parallel circuit is the sum of the currents through the individual paths* ($It = I1 + I2 + I3$). Notice in figure 4-4 that each of the resistors is connected directly across the incoming power line. Therefore, *each component in a parallel circuit has the same voltage drop*.

Each time a new component is added to a parallel circuit, a new path for current flow is created. Since there is less opposition to current flow each time a component is added, the total resistance of the circuit is decreased. The total resistance of a parallel circuit can be found using either of two formulas. The first of these formulas is:

$$Rt = \frac{R1 \times R2}{R1 + R2}$$

The second formula is:

$$\frac{1}{Rt} = \frac{1}{R1} + \frac{1}{R2} + \frac{1}{R3}$$

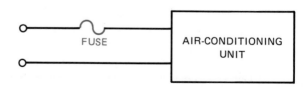

FIGURE 4-3 Fuses are connected in series.

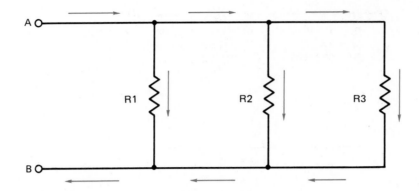

FIGURE 4-4 Parallel circuits have more than one path for current flow.

Figure 4-5 shows a parallel circuit containing three resistors with the values of 15, 10, and 30 ohms. The total resistance of the circuit can be found by using either of the two formulas.

$$Rt = \frac{15 \times 10}{15 + 10} = \frac{150}{25} = 6$$

Notice that in this formula only two of the resistors can be found at a time. It is now necessary to use the total resistance of the first two resistors and use that value for R1 in the formula. Resistor R3 is used in the R2 position in the formula.

$$Rt = \frac{6 \times 30}{6 + 30}$$

$$Rt = \frac{180}{36}$$

$$Rt = 5 \text{ ohms}$$

The total resistance of this parallel circuit is 5 ohms. The second formula can be used to find the total resistance by plugging in the values of resistance into the formula. When this is done, it becomes a matter of adding fractions. When fractions are to be added, the first thing that must be done is to find some number all the denominators will divide into. This is called finding a common denominator. For this problem, 30 will be the common denominator.

$$\frac{1}{Rt} = \frac{1}{15} + \frac{1}{10} + \frac{1}{30}$$

$$\frac{1}{Rt} = \frac{2}{30} + \frac{3}{30} + \frac{1}{30}$$

$$\frac{1}{Rt} = \frac{6}{30} \quad \frac{Rt}{1} = \frac{30}{6}$$

$$Rt = 5 \text{ ohms}$$

Notice that both formulas give the same answer.
    Another method of finding the total resistance

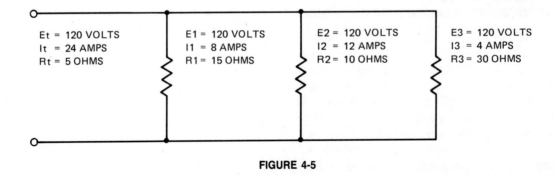

Et = 120 VOLTS   E1 = 120 VOLTS   E2 = 120 VOLTS   E3 = 120 VOLTS
It = 24 AMPS   I1 = 8 AMPS   I2 = 12 AMPS   I3 = 4 AMPS
Rt = 5 OHMS   R1 = 15 OHMS   R2 = 10 OHMS   R3 = 30 OHMS

FIGURE 4-5

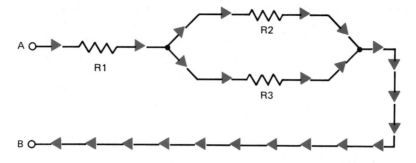

FIGURE 4-6 Combination circuit

in a parallel circuit is to find the reciprocal of each individual resistor. A third rule for parallel circuits states that *total resistance is the reciprocal of the sum of the reciprocals of the individual resistors.* The problem can, therefore, be solved by finding the reciprocal of each individual resistor, adding them together, and finding the reciprocal of the sum. (The reciprocal of any number can be found by dividing that number into 1.) The problem can be solved as follows:

$$\frac{1}{Rt} = \frac{1}{R1} + \frac{1}{R2} + \frac{1}{R3}$$

$$\frac{1}{Rt} = \frac{1}{15} + \frac{1}{10} + \frac{1}{30}$$

$$\frac{1}{Rt} = .0667 + .1 + .0333$$

$$\frac{1}{Rt} = .02$$

$$Rt = 5 \text{ ohms}$$

If 120 volts is applied to the circuit, the values of voltage and current for the entire circuit can be found. Since each of the resistors is connected directly across the power line, each resistor will have the same voltage drop of 120 volts. The current flow through resistor R1 can be found using Ohm's Law.

$$I = \frac{E}{R}$$

$$I = \frac{120}{15}$$

$$I = 8 \text{ amps}$$

The current flow through resistor R2 is (120/10 = 12 amps). The current flow through resistor R3 is (120/30 = 4 amps). The total current flow in the circuit can be found by using the formula

$$It = \frac{Et}{Rt}$$

$$It = \frac{120}{5}$$

$$It = 24 \text{ amps}$$

Notice that the total current can also be found by adding the currents flowing through the individual resistors (8 + 12 + 4 = 24 amps).

Most circuits are connected in parallel. The lights and outlets in a house are connected in parallel. Since all of the lights and outlets are connected in parallel, each light has an applied voltage of 120 volts, and each outlet can supply 120 volts to whatever is connected to it. If the lights in a house were wired in series, all of the lights would have to be turned on before any of them would burn.

COMBINATION CIRCUITS

Combination circuits are seldom, if ever, encountered by the air-conditioning service technician in the field. For this reason, only a brief description of them is given. A combination circuit contains both series and parallel connections within the same circuit. In figure 4-6, resistor R1 is connected in series with resistors R2 and R3. Resistors R2 and R3 are connected in parallel with each other. If it is assumed that current flows from point "A" to point "B," all of the current would have to flow through resistor R1. At the junction of resistors R2 and R3, however, the current divides and flows through separate paths. The amount of current that flows through each resistor is determined by their resistance. Notice that all of the circuit current must flow through R1. Since there is only one current path through R1, it is connected in series with the rest of the circuit. When the current reaches the junction of resistors R2 and R3, there is more than one path for current flow. These resistors are, therefore, connected in parallel. Combination circuits are encountered in electronic circuits, but seldom found in power wiring such as that used in industry or the home.

REVIEW QUESTIONS

1. List the three basic types of electrical circuits.

2. What is the major characteristic of a series circuit?

3. List the three basic rules for series circuits.

4. What is the major characteristic of a parallel circuit?

5. List the three basic rules for the parallel circuit.

6. What type of circuit is used most often in industry and the home?

7. What type of circuit is used the least in industry and the home?

8. Three resistors valued at 300 ohms, 200 ohms, and 600 ohms are connected in series. What is their total resistance?

9. If the three resistors in question #8 are connected in parallel, what is their total resistance?

10. How are fuses and circuit breakers connected in a circuit and why are they connected this way?

UNIT 5
Electrical Services

Air-conditioning equipment must be connected to an electrical service. The type of air-conditioning equipment used is generally determined by the type of electrical service available to operate it. Therefore, the air-conditioning technician must have knowledge of different types of electrical services.

POWER GENERATION

In the United States and Canada, power is generated as a three-phase 60-Hertz voltage. The term Hertz means 60 cycles per second. This means that the voltage increases from zero to its maximum positive value, returns to zero, increases to its maximum negative value and returns to zero 60

times each second. Figure 5-1 shows one complete cycle of AC voltage.

The term *three phase* means that there are three separate voltage waveforms produced by the alternator. An alternator is a generator that produces AC voltage. For the alternator to produce the three phases, the internal windings of the alternator—called the stator—are wound 120° apart. Figure 5-2 illustrates the windings of an alternator. The moving part of the alternator, called the rotor, is actually a large electromagnet. When the magnet is turned, the magnetic field cuts through the windings of the stator and induces a voltage into them. The amount of voltage induced is controlled by the strength of the magnetic field, and the frequency or hertz is controlled by the speed of rotation of the magnet. Since the windings of the stator are

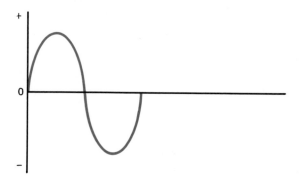

FIGURE 5-1 AC sine wave

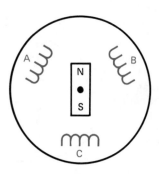

FIGURE 5-2 The windings of an alternator are 120° apart.

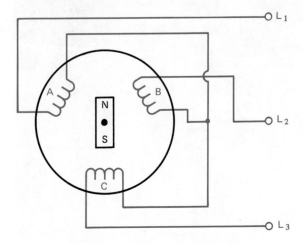

FIGURE 5-3 Wye or star connection

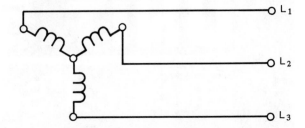

FIGURE 5-4 Schematic of a wye connection

physically wound 120° apart, the three voltages are 120° out of phase with each other. The windings of the stator are connected to form one of the two basic three-phase connections. These connections are the *delta* and *wye*.

WYE CONNECTION

The wye connection is also referred to as the *star* connection. This connection is made by joining one end of each of the windings together as shown in figure 5-3. The connection shown in fig-

ure 5-4 is a wye connection that has been drawn schematically to make it easier to see and understand. Notice how one end of each of the windings is joined at the centerpoint. The wye connection can be used to provide an increase in the output or line voltage. The phase voltage is the voltage produced across one of the windings. The line voltage is the voltage produced across the output points of the connection. Figure 5-5 shows a wye connection connected to a three-phase load bank. Ammeters and voltmeters are used to illustrate the differences between phase values and line values. Notice that the phase value of voltage is measured from the output of the winding, at point "C," to the centerpoint of the wye connection at the point labeled "O." The line value is measured across two of the output points of the connection (B&C). The phase current meter is inserted in the winding of the alternator, and the line current meter is inserted in

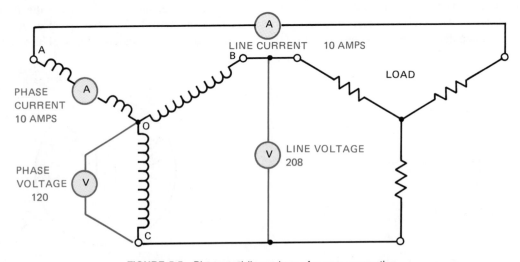

FIGURE 5-5 Phase and line values of a wye connection

the output line. Notice also that the two ammeters are indicating the same value of current. *In a wye connection, phase current and line current are equal.* The voltages, however, are not. *The line voltage in a wye connection is 1.732 times greater than the phase voltage* (1.732 is the square root of 3). The reason for this voltage increase is because the voltages are 120° out of phase with each other. Figure 5-6 shows a diagram to illustrate this. Since the three voltages are out of phase with each other, they will be added. Vector addition must be used, however, because of the 120° phase shift. If three voltages are shown in a length that corresponds to 120 volts, and a resultant is drawn to the point of intersection, it will be found that the length of the resultant corresponds to 208 volts. The 120° phase shift between voltages is the reason the two 120-volt phases add to produce 208 volts instead of 240 volts.

Wye-connected systems often use a fourth conductor connected to the center of the connection. This conductor becomes the neutral, figure 5-7. Notice in this connection that the voltage be-

tween any line and neutral is the phase voltage or 120 volts, and the voltage between any two of the lines is 208 volts. The 208/120-volt three-phase connection is very common in industry and commercial buildings. Another very common three-phase four-wire connection is shown in figure 5-8. This is a 480/277-volt connection. Two hundred seventy-seven volts is often used in large stores and office buildings to operate the fluorescent lights while the 480-volt is used to operate large air-conditioning systems. The 120-volt connections are provided by transformers that step down the 480 volts to 120 volts.

DELTA CONNECTION

The next connection to be covered is the delta. A schematic diagram of a delta connection is shown in figure 5-9. This connection gets its name from the fact that it looks like the Greek letter delta (Δ). Figure 5-10 shows a delta system connected to a three-phase load bank. Ammeters and voltmeters are used to illustrate the differences in phase and line values of voltage and current. Notice that the values of phase voltage and line voltage are equal for the delta connection. One of the rules for three-phase systems states that *line voltage and phase voltage are equal in a delta connection.* The ammeters, however, are not equal. *In a delta connection, the line current is 1.732 times greater than the phase current.* This is the reason that the delta connection is so popular in industry. The current flow through the windings of a transformer are less than the line amps if the transformer bank is connected in delta.

HIGH-LEG SYSTEM

Figure 5-11 illustrates another common type of transformer connection. This is a 240/120-volt system with a high leg. Three transformers are connected to form a delta connection. One of the transformers is larger than the other two, however, and is center tapped. The large transformer must be able to supply power for both three-phase and single-phase loads. The other two transformers supply power for the three-phase loads only. If the

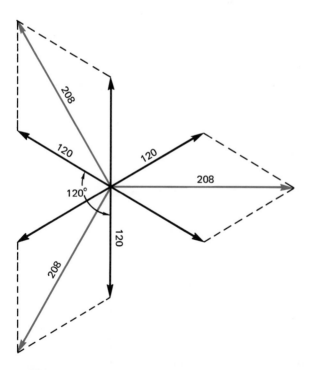

FIGURE 5-6 Vector diagram of the phase and line values in a three-phase system

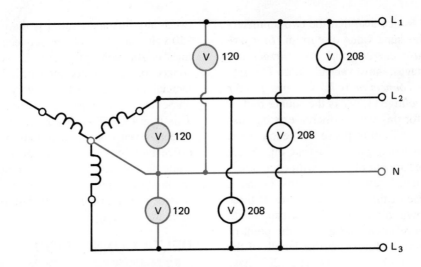

FIGURE 5-7 Fourth wire connected for a neutral

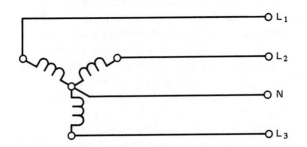

FIGURE 5-8 A 480/277-volt connection

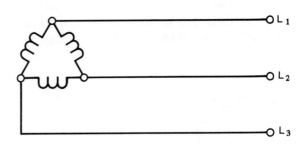

FIGURE 5-9 The delta connection

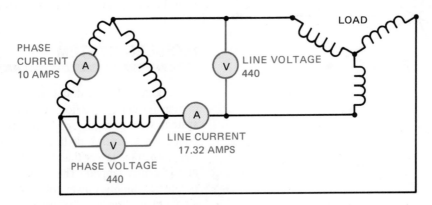

FIGURE 5-10

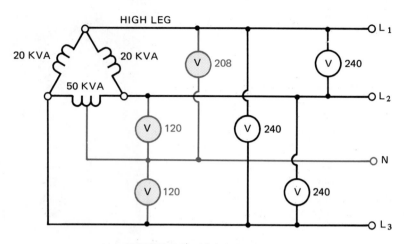

FIGURE 5-11   High leg system

phase voltage of the transformers is 240 volts, the voltage between any of the three lines is 240 volts. If the center-tap connection is used as a neutral conductor, however, the voltages between L2 and neutral, and L3 and neutral will be 120 volts. Therefore, L2, L3, and neutral are used to supply 240/120 volts for single-phase loads. Care must be taken not to connect a 120-volt device across L1 and neutral. Line L1 is known as a high leg and a voltage of about 208 volts exists between these two points.

OPEN-DELTA SYSTEM

Another type of three-phase service is known as the open delta. The open-delta system has the advantage of needing only two transformers to pro-

vide three-phase voltage. This connection is often used when the amount of three-phase power needed is low, or if the power needs are expected to increase in the future. The open-delta connection, however, does have some disadvantages. The total output power is only 84% of the combined rating of the transformers. If the two transformers shown in figure 5-12 each have a power rating of 25 KVA (kilovolt amps), the total delivered power of this connection is only 42 KVA (25 + 25 = 50) (50 × 84% = 42). If at a later date the power require-

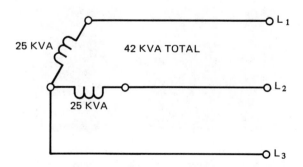

FIGURE 5-12   Open delta system

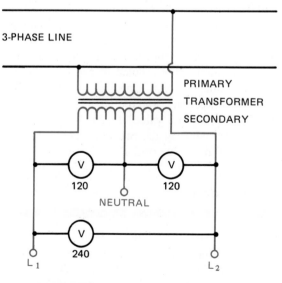

FIGURE 5-13   Single-phase transformer

FIGURE 5-14 The voltage across a single transformer winding is 180° out of phase.

ments increase, a third transformer can be added to close the delta. The total output power of this connection is the combined rating of all three transformers. In this case it will be 75 KVA (25 + 25 + 25 = 75).

SINGLE-PHASE SERVICE

A single-phase 240/120-volt system is formed by connecting a single transformer to a three-phase line. The primary of the transformer shown in figure 5-13 is connected to the three-phase lines of the power company. The secondary voltage of the transformer is 240 volts. The secondary winding of the transformer is center tapped, and this connection becomes the neutral conductor. If the voltage across the entire secondary is measured, it will be 240 volts. If the voltage between either of the secondary leads is measured to the center tap, it will be 120 volts. The reason this is true for single phase and not three phase is because the voltage across a single transformer winding is 180° out of phase. Figure 5-14 illustrates this condition. If a vector diagram is constructed, it can be seen that the two voltages are in opposite directions. The distance between the centerpoint and either voltage will correspond to a value of 120. If the distance is measured between the two outer points, it will correspond to a value of 240.

PANEL BOX

Regardless of the type of service used, connection will be made at a fuse or circuit-breaker box. Figure 5-15 shows a 150-amp single-phase circuit-breaker panel. Circuit breakers are made in

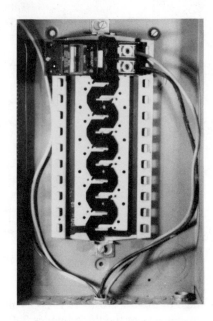

FIGURE 5-15   150 amp single-phase panel

different sizes and types. Figure 5-16 shows three different types of circuit breakers. The single-pole breaker is used for connecting a 120-volt circuit, the two-pole breaker is used for connecting a 240-volt single-phase circuit, and the three-pole breaker is used for connecting a three-phase circuit. The three-pole breaker must be used with a three-phase circuit-breaker panel and cannot be used in a single-phase panel.

When a 120-volt connection is to be made, cable is brought into the panel. A two-conductor romex cable contains three wires—a black, white, and bare copper. The bare copper wire is the

FIGURE 5-16   Single-pole, double-pole, three-pole circuit breakers

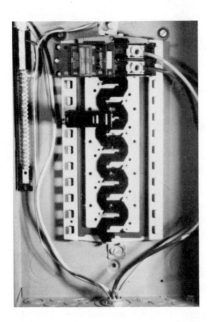

FIGURE 5-17 120-volt single-phase connection

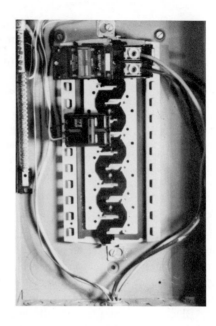

FIGURE 5-19 240-volt single-phase three-wire connection

grounding wire or safety wire and is not considered a circuit conductor. Only the black and white wires are considered to be circuit conductors. The black wire is used as the "hot" conductor and the white wire is used as the neutral. Figure 5-17 shows a 120 single-phase circuit connected into the panel

box. Notice that the black wire is connected to the circuit breaker, and the white wire is connected to the neutral buss. Notice also that the bare copper wire is connected to the neutral buss with the white wire.

When a 240-volt connection must be made, a two-pole circuit breaker is used. If the connection is to use only two-circuit conductors as shown in

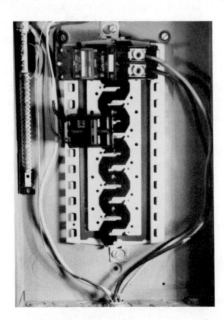

FIGURE 5-18 240-volt single-phase two-wire connection

FIGURE 5-20 Replaceable-link type of fuse

240-6. Standard Ampere Ratings. The standard ampere ratings for fuses and inverse time circuit breakers shall be considered 15, 20, 25, 30, 35, 40, 45, 50, 60, 70, 80, 90, 100, 110, 125, 150, 175, 200, 225, 250, 300, 350, 400, 450, 500, 600, 700, 800, 1000, 1200, 1600, 2000, 2500, 3000, 4000, 5000, and 6000.

Exception: Additional standard ratings for fuses shall be considered 1, 3, 6, 10, and 601.

FIGURE 5-21 Standard fuse ratings (Reprinted with permission from NFPA 70, *National Electrical Code®*, Copyright © 1989, National Fire Protection Association, Quincy, Massachusetts 02269. This reprinted material is not the complete and official position of the NFPA on the referenced subject, which is represented only by the standard in its entirety. National Electrical Code® and NEO® are registered trademarks of the National Fire Protection Association, Inc., Quincy, MA.)

figure 5-18, the black wire connects to one pole of the two-pole breaker. The National Electrical Code does not permit a white wire to be used as a hot circuit conductor. For this reason the wire must be identified by wrapping a piece of colored tape around it. The tape can be any color except white, gray, or green. Black or red tape is generally used. The identified conductor is then connected to the other pole of the two-pole breaker. The bare copper wire is connected to the neutral buss.

If a 240-volt three-wire circuit is to be connected to the panel, a three-conductor cable is used. The three-conductor cable contains four wires—a black, red, white, and green. The green is the grounding or safety wire and is not considered a circuit conductor. Figure 5-19 shows a 240-volt three-wire connection. The black and red wires are connected to the two poles of the circuit breaker. The white and green wires are connected to the neutral buss.

When a three-phase panel connection is made, a three-pole circuit breaker is used. There may or may not be a neutral depending on the type of circuit. For example, a 208/120-volt connection would use a fourth wire connected to the neutral buss. A 440-volt straight, three-phase connection would use

only three conductors connected to a three-pole breaker.

FUSES

Circuit breakers are not the only means used to provide circuit protection. Fuses are still used to a great extent. Fuses are rated in two ways—by voltage and current. The voltage rating of a fuse indicates the amount of voltage the fuse is designed to interrupt without arcing across. Although fuses can be obtained that have ratings of several thousand volts, the most common fuses used in the air-conditioning field are 300 volt and 600 volt. The 600-volt fuse is longer to provide a greater distance between the two contact ends if the fuse link should blow. The extra length is needed at higher voltages to prevent arc-over.

Figure 5-20 shows a type of fuse that uses a replaceable link. When this fuse blows, the fuse cartridge can be taken apart and the fuse link replaced. This type of fuse is more expensive to purchase, but it could be a savings if the fuse has to be replaced frequently.

Fuses used for circuit protection are made in

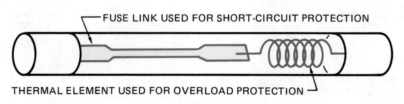

FUSE LINK USED FOR SHORT-CIRCUIT PROTECTION

THERMAL ELEMENT USED FOR OVERLOAD PROTECTION

FIGURE 5-22 Dual element fuse

FIGURE 5-23 Three-phase fused disconnect

standard ampere ratings. Figure 5-21 shows these ratings as taken from the national electrical code. Fuses for air-conditioning and refrigeration equipment are normally sized at 175% of the rated full-load current of the motor. Under some conditions, however, compressors can be fused as much as 225% of their full-load running current. If the fuse size needed does not correspond with one of the standard fuse sizes, the next smaller size fuse will have to be used. For example, assume it has been determined that a fuse rating of 130 amps is needed. The standard ratings chart for fuses shown in figure 5-21 does not list a 130-amp fuse. Therefore, the closest standard rating less than 130 amps is 125 amps. A 125-amp fuse will be used. Notice that fuses can be sized as much as 225% of the full-load current of the compressor. The reason that fuses are sized this much above the running current of the motor is to permit the fuse the ability to withstand the starting current of the motor. Fuses are designed to protect the circuit against short circuits, and not used to protect the motor from overloads. Overload protection for the motor is provided by the overload relay, which will be covered in a later unit, or by dual-element fuses. Dual-ele-

ment fuses are designed to provide both types of protection. Figure 5-22 illustrates a dual-element fuse.

FUSED DISCONNECTS

Fused disconnects provide both a disconnect switch and fuse holders. Figure 5-23 shows a fused disconnect used for three-phase circuits. Fused disconnects, like fuses, have standard ratings. The standard sizes for fused disconnects are shown in figure 5-24. The rating of the disconnect indicates the maximum size of fuse that can be used in that enclosure. For example, assume the 125-amp fuse discussed earlier in this unit is to be mounted in a disconnect. Since the fuse size is greater than 100 amps, it cannot be mounted in a 100-amp enclosure. The next standard size enclosure is 200 amps. The 125-amp fuses will have to be mounted in a 200-amp disconnect.

When servicing equipment, it is often necessary to turn off the power to the equipment. When this is necessary, certain precautions should be taken by the service technician. Remember that your life is your own and do not trust someone else not to turn the circuit back on while it is being serviced. Most industries provide a tag that is hung on the

Standard disconnect sizes
30 amp
60 amp
100 amp
200 amp
400 amp
600 amp
1000 amp
1200 amp
1600 amp
2000 amp
3000 amp
4000 amp
5000 amp
6000 amp

FIGURE 5-24

disconnect while it is being serviced. A paper tag, however, cannot stop someone from turning the power back on. For this reason, a small padlock should be used to lock the disconnect in the off position. If a lock is not available, the fuses should be removed with fuse pullers. There is no such thing as being too safe when working with high-voltage electricity.

REVIEW QUESTIONS

1. What is an alternator?

2. What controls the output voltage of an alternator?

3. What controls the frequency of the alternator?

4. How many degrees out of phase with each other are the voltages of a three-phase system?

5. What are the two major types of three-phase connections?

6. List the rules concerning line and phase values of current and voltage in a wye connection.

7. List the rules concerning line and phase values of current and voltage in a delta connection.

8. In a high-leg delta-connected system, what is the voltage between the high leg and neutral?

9. What type of three-phase transformer connection uses only two transformers?

10. How many degrees out of phase are the voltages of a single-phase system?

11. A two-conductor romex cable contains three wires. Which wire is not counted and why?

12. What type of circuit breaker is used to make a 240-volt connection?

13. Where does the grounding conductor connect in a panel?

14. In what two electrical units are fuses rated?

15. It has been calculated that a 290-amp fuse is needed to protect the circuit supplying an air-conditioning compressor. What standard rating of fuse should be used?

16. What size fuse disconnect will be used for the fuse in question #15?

17. What is a dual-element fuse?

UNIT 6
Wire Size and Voltage Drop

When installing air-conditioning equipment it is important to use the proper size wire. If wire is used that is larger than needed, it is an unnecessary expense. If wire is used that is too small, it will cause excessive voltage drop and damage the equipment.

WIRE RESISTANCE

Most people think of wire as having zero resistance. In fact, many electrical calculations are made that assume the resistance of the wire is so little that it is negligible. In actual practice, however, all wire has resistance. There are four factors that determine the resistance of a piece of wire. These factors are:

1. The diameter of the wire,
2. The material the wire is made of,
3. The length of the wire,
4. The temperature of the wire.

AREA

The cross-sectional area of wire is measured in *circular mils*. The circular mil area of a wire can be found by finding the diameter of the wire in thousandths of an inch (1 mil = 1/1000 inch) and squaring that number. (Squaring a number means to multiply that number by itself.) For ex-

ample, assume a piece of wire is measured with a micrometer and is found to have a diameter of 8 thousandths of an inch (.008). The circular mil area of the wire is 64 (8 × 8 = 64). Notice that 64 is written as a whole number, not a decimal number. A wire that has a diameter of .064 inch has a circular mil area of 4096 (64 × 64 = 4096).

The circular mil area of stranded wire is determined by finding the area of one of the strands and then multiplying by the number of strands. For example, assume that a wire has 24 strands of wire that are .012 inches in diameter. The area of one wire is 144 CM (circular mils). The entire conductor has a circular mil area of 3456 CM (144 × 24 = 3456). Large wire is generally stranded to make it easier to bend.

The larger the diameter of a wire, the less resistance it will have and the more current it can carry, figure 6-1. Current flowing through a wire is very similar to water flowing through a pipe. A large wire can carry more current at a specific voltage than a small wire. A large pipe can carry more water at a specific pressure than a small pipe.

MATERIAL

A standard measurement used for finding the resistance of wire is the *mil-foot*. A mil-foot of wire is a piece of wire one circular mil in diameter and one foot long. If the resistance of a mil-foot of dif-

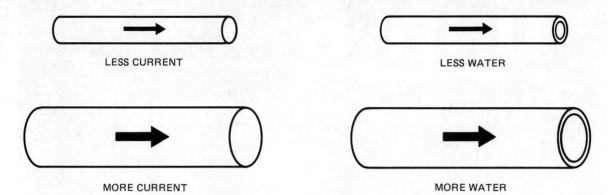

FIGURE 6-1 Larger wire can carry more current.

ferent types of wire is found, a mathematical formula can be used to determine the resistance of different types, sizes, and lengths of wire. This formula is:

$$R = K \times L / CM$$

Where: K = the ohms per mil-foot of the wire.
L = the length of the wire in feet.
CM = the circular mil area of the wire.

The table in figure 6-2 gives the resistance of different types of wire in ohms per mil-foot. Using the table shown in figure 6-3, the diameter and circular mil area for different sizes of wire can be found.

Problem #1: What is the resistance of a piece of #18 AWG (American Wire Gauge) copper wire 400 feet long?

K = OHMS RESISTANCE PER MIL-FOOT (AT 70° F)	
Aluminum	17
Brass	42
Cadmium Bronze	12
Copper	10.4
Copperclad Aluminum (20% Cu)	15.2
Copperweld	26–34
Iron	60
Nichrome	600
Silver	9.6
Steel	75
Tungsten	33

FIGURE 6-2

Solution: First, state the formula to be used.

$$R = \frac{K \times L}{CM}$$

Second, substitute known numeric values in the formula. The value of "K" can be found in the table shown in figure 6-2. The "K" value for copper is 10.4. The CM area of #18 AWG wire can be found from the chart in figure 6-3. The circular mil area of #18 AWG wire is 1624 CM. If these values are substituted in the formula for letters, the formula will be:

$$R = \frac{10.4 \times 400}{1624}$$

$$R = \frac{4160}{1624}$$

$$R = 2.56 \text{ ohms}$$

Problem #2: What is the resistance of a piece of #12 AWG aluminum wire 250 feet long?

Solution:

$$R = \frac{K \times L}{CM}$$

$$R = \frac{17 \times 250}{6530}$$

$$R = \frac{4250}{6530}$$

$$R = .6508 \text{ ohms}$$

TABLE A-2 AMERICAN WIRE GAUGE TABLE

B & S Gauge No.	Diam. in Mils	Area in Circular Mils	Ohms per 1 000 Ft. (ohms per 100 m)			Pounds per 1 000 Ft. (kg per 100 m)	
			Copper* 68°F (20°C)	Copper* 167°F (75°C)	Aluminum 68°F (20°C)	Copper	Aluminum
0000	460	211 600	.049 (.016)	.0596 (.0195)	.0804 (.0263)	640 (95.2)	195 (29.0)
000	410	167 800	.0618 (.020)	.0752 (.0246)	.101 (.033)	508 (75.5)	154 (22.9)
00	365	133 100	.078 (.026)	.0948 (.031)	.128 (.042)	403 (59.9)	122 (18.1)
0	325	105 500	.0983 (.032)	.1195 (.0392)	.161 (.053)	320 (47.6)	97 (14.4)
1	289	83 690	.1239 (.0406)	.151 (.049)	.203 (.066)	253 (37.6)	76.9 (11.4)
2	258	66 370	.1563 (.0512)	.190 (.062)	.526 (.084)	201 (29.9)	61.0 (9.07)
3	229	52 640	.1970 (.0646)	.240 (.079)	.323 (.106)	159 (23.6)	48.4 (7.20)
4	204	41 740	.2485 (.0815)	.302 (.099)	.408 (.134)	126 (18.7)	38.4' (5.71)
5	182	33 100	.3133 (.1027)	.381 (.125)	.514 (.168)	100 (14.9)	30.4 (4.52)
6	162	26 250	.395 (1.29)	.481 (.158)	.648 (.212)	79.5 (11.8)	24.1 (3.58)
7	144	20 820	.498 (.163)	.606 (.199)	.817 (.268)	63.0 (9.37)	19.1 (2.84)
8	128	16 510	.628 (.206)	.764 (.250)	1.03 (.338)	50.0 (7.43)	15.2 (2.26)
9	114	13 090	.792 (.260)	.963 (.316)	1.30 (.426)	39.6 (5.89)	12.0 (1.78)
10	102	10 380	.999 (.327)	1.215 (.398)	1.64 (.538)	31.4 (4.67)	9.55 (1.42)
11	91	8 234	1.260 (.413)	1.532 (.502)	2.07 (.678)	24.9 (3.70)	7.57 (1.13)
12	81	6 530	1.588 (.520)	1.931 (.633)	2.61 (.856)	19.8 (2.94)	6.00 (.89)
13	72	5 178	2.003 (.657)	2.44 (.80)	3.29 (1.08)	15.7 (2.33)	4.8 (.71)
14	64	4 107	2.525 (.828)	3.07 (1.01)	4.14 (1.36)	12.4 (1.84)	3.8 (.56)
15	57	3 257	3.184 (1.043)	3.98 (1.27)	5.22 (1.71)	9.86 (1.47)	3.0 (.45)
16	51	2 583	4.016 (1.316)	4.88 (1.60)	6.59 (2.16)	7.82 (1.16)	2.4 (.36)
17	45.3	2 048	5.06 (1.66)	6.16 (2.02)	8.31 (2.72)	6.20 (.922)	1.9 (.28)
18	40.3	1 624	6.39 (2.09)	7.77 (2.55)	10.5 (3.44)	4.92 (.731)	1.5 (.22)
19	35.9	1 288	8.05 (2.64)	9.79 (3.21)	13.2 (4.33)	3.90 (.580)	1.2 (.18)
20	32.0	1 022	10.15 (3.33)	12.35 (4.05)	16.7 (5.47)	3.09 (.459)	0.94 (.14)
21	28.5	810	12.8 (4.2)	15.6 (5.11)	21.0 (6.88)	2.45 (.364)	.745 (.110)
22	25.4	642	16.1 (5.3)	19.6 (6.42)	26.5 (8.69)	1.95 (.290)	.591 (.09)
23	22.6	510	20.4 (6.7)	24.8 (8.13)	33.4 (10.9)	1.54 (.229)	.468 (.07)
24	20.1	404	25.7 (8.4)	31.2 (10.2)	42.1 (13.8)	1.22 (.181)	.371 (.05)
25	17.9	320	32.4 (10.6)	39.4 (12.9)	53.1 (17.4)	0.97 (.14)	.295 (.04)
26	15.9	254	40.8 (13.4)	49.6 (16.3)	67.0 (22.0)	.77 (.11)	.234 (.03)
27	14.2	202	51.5 (16.9)	62.6 (20.5)	84.4 (27.7)	.61 (.09)	.185 (.03)
28	12.6	160	64.9 (21.3)	78.9 (25.9)	106 (34.7)	.48 (.07)	.147 (.02)
29	11.3	126.7	81.8 (26.8)	99.5 (32.6)	134 (43.9)	.384 (.06)	.117 (.02)
30	10.0	100.5	103.2 (33.8)	125.5 (41.1)	169 (55.4)	.304 (.04)	.092 (.01)
31	8.93	79.7	130.1 (42.6)	158.2 (51.9)	213 (69.8)	.241 (.04)	.073 (.01)
32	7.95	63.2	164.1 (53.8)	199.5 (65.4)	269 (88.2)	.191 (.03)	.058 (.01)
33	7.08	50.1	207 (68)	252 (82.6)	339 (111)	.152 (.02)	.046 (.01)
34	6.31	39.8	261 (86)	317 (104)	428 (140)	.120 (.02)	.037 (.01)
35	5.62	31.5	329 (108)	400 (131)	540 (177)	.095 (.01)	.029
36	5.00	25.0	415 (136)	505 (165)	681 (223)	.076 (.01)	.023
37	4.45	19.8	523 (171)	636 (208)	858 (281)	.0600 (.01)	.0182
38	3.96	15.7	660 (216)	802 (263)	1080 (354)	.0476 (.01)	.0145
39	3.53	12.5	832 (273)	1012 (332)	1360 (446)	.0377 (.01)	.0115
40	3.15	9.9	1049 (344)	1276 (418)	1720 (564)	.0299 (.01)	.0091
41							
42	2.50	6.3					
43							
44	1.97	3.9					

*Resistance figures are given for standard annealed copper. For hard-drawn copper add 2%

FIGURE 6-3 (Reprinted, with permission, from Loper, *Direct Current Fundamentals*, Table A-2)

Table 310-16.   Ampacities of Insulated Conductors
Rated 0-2000 Volts, 60° to 90°C (140° to 194°F)
Not More Than Three Conductors in Raceway or Cable or Earth
(Directly Buried), Based on Ambient Temperature of 30°C (86°F)

Size	Temperature Rating of Conductor. See Table 310-13.								Size
	60°C (140°F)	75°C (167°F)	85°C (185°F)	90°C (194°F)	60°C (140°F)	75°C (167°F)	85°C (185°F)	90°C (194°F)	
AWG kcmil	TYPES †TW, †UF	TYPES †FEPW, †RH, †RHW, †THHW, †THW, †THWN, †XHHW, †USE, †ZW	TYPE V	TYPES TA, TBS, SA SIS, †FEP, †FEPB, †RHH, †THHN, †THHW, †XHHW	TYPES †TW, †UF	TYPES †RH, †RHW, †THHW, †THW, †THWN, †XHHW †USE	TYPE V	TYPES TA, TBS, SA, SIS, †RHH, †THHW, †THHN, †XHHW	AWG kcmil
	COPPER				ALUMINUM OR COPPER-CLAD ALUMINUM				
18	....	....	....	14	....	....	....	....	
16	....	....	18	18	....	....	....	....	
14	20†	20†	25	25†	....	....	....	....	
12	25†	25†	30	30†	20†	20†	25	25†	12
10	30	35†	40	40†	25	30†	30	35†	10
8	40	50	55	55	30	40	40	45	8
6	55	65	70	75	40	50	55	60	6
4	70	85	95	95	55	65	75	75	4
3	85	100	110	110	65	75	85	85	3
2	95	115	125	130	75	90	100	100	2
1	110	130	145	150	85	100	110	115	1
1/0	125	150	165	170	100	120	130	135	1/0
2/0	145	175	190	195	115	135	145	150	2/0
3/0	165	200	215	225	130	155	170	175	3/0
4/0	195	230	250	260	150	180	195	205	4/0
250	215	255	275	290	170	205	220	230	250
300	240	285	310	320	190	230	250	255	300
350	260	310	340	350	210	250	270	280	350
400	280	335	365	380	225	270	295	305	400
500	320	380	415	430	260	310	335	350	500
600	355	420	460	475	285	340	370	385	600
700	385	460	500	520	310	375	405	420	700
750	400	475	515	535	320	385	420	435	750
800	410	490	535	555	330	395	430	450	800
900	435	520	565	585	355	425	465	480	900
1000	455	545	590	615	375	445	485	500	1000
1250	495	590	640	665	405	485	525	545	1250
1500	520	625	680	705	435	520	565	585	1500
1750	545	650	705	735	455	545	595	615	1750
2000	560	665	725	750	470	560	610	630	2000

AMPACITY CORRECTION FACTORS

Ambient Temp. °C	For ambient temperatures other than 30°C (86°F), multiply the ampacities shown above by the appropriate factor shown below.								Ambient Temp. °F
21-25	1.08	1.05	1.04	1.04	1.08	1.05	1.04	1.04	70-77
26-30	1.00	1.00	1.00	1.00	1.00	1.00	1.00	1.00	79-86
31-35	.91	.94	.95	.96	.91	.94	.95	.96	88-95
36-40	.82	.88	.90	.91	.82	.88	.90	.91	97-104
41-45	.71	.82	.85	.87	.71	.82	.85	.87	106-113
46-50	.58	.75	.80	.82	.58	.75	.80	.82	115-122
51-55	.41	.67	.74	.76	.41	.67	.74	.76	124-131
56-60	....	.58	.67	.71	....	.58	.67	.71	133-140
61-70	....	.33	.52	.58	....	.33	.52	.58	142-158
71-80	....	....	.30	.41	....	....	.30	.41	160-176

† Unless otherwise specifically permitted elsewhere in this Code, the overcurrent protection for conductor types marked with an obelisk (†) shall not exceed 15 amperes for 14 AWG, 20 amperes for 12 AWG, and 30 amperes for 10 AWG copper; or 15 amperes for 12 AWG and 25 amperes for 10 AWG aluminum and copper-clad aluminum after any correction factors for ambient temperature and number of conductors have been applied.

FIGURE 6-4 (Reprinted with permission from NFPA 70, *National Electrical Code®*, Copyright © 1989, National Fire Protection Association, Quincy, Massachusetts 02269. This reprinted material is not the complete and official position of the NFPA on the referenced subject, which is represented only by the standard in its entirety.)

TEMPERATURE

The resistance of a piece of wire is also affected by temperature. As the temperature increases, the resistance of wire increases also. Notice that the charts in figures 6-2 and 6-3 state the resistance of wire at a specific temperature. Most wire tables will provide some means for determining the resistance of wire as temperature increases.

Figure 6-4 shows table 310-16 of the national electrical code. Notice that at the bottom of the table ampacity correction factors are given.

INSULATION

The type of insulation around the wire also partly determines the amount of current the wire is permitted to carry. Some types of insulation can withstand more heat than others, and are, therefore, permitted to carry more current. For example, in table 310-16, the type of wire insulation is listed at the top of each column. Notice the different temperature ratings for the types of insulation listed. Also notice the amount of current a wire is permitted to carry for different types of insulation. Find a #2 AWG conductor on the far left-hand side of the wire table. Notice the different amounts of current this conductor is permitted to carry with different types of insulation.

VOLTAGE RATING

Wire also has a voltage rating. The voltage rating of wire has nothing to do with the type of material the wire is made of or its diameter. The voltage rating is determined by the type of insulation. Most wire used in industry has a voltage rating of 600 volts. The amount of voltage the insulation can effectively hold off is determined by the material the insulation is made of and its thickness.

SIZING THE WIRE

The wire size used to connect an air-conditioning unit is generally determined by two factors. These are the amount of current the unit will draw when operating, and the distance the wire must be run. For example, assume an air-conditioning unit is to have a current draw of 24 amps, and is to be located 30 feet from the circuit-breaker panel. Also assume this unit is to be connected to a 30-amp circuit breaker, and that the type of wire insulation is THWN. The proper wire size can be chosen from the table shown in figure 6-4. First locate the type of wire insulation used. Type THWN insulation can be found in the third column. Read down this column until the nearest amperage is found. 35 amps will be used since it is to be connected to a 30-amp circuit breaker. This corresponds to a #10 AWG copper wire. To determine the amount of voltage drop this wire will have, the resistance of the wire must first be found.

$$R = \frac{K \times L}{CM}$$

$$R = \frac{10.4 \times 60}{10,380}$$ (60 feet is used because the two-wire cable is 30 feet long. This means there is 30 feet of wire going to the unit and 30 feet of wire returning to the panel)

$$R = .060 \text{ ohms}$$

The amount of voltage dropped by the wire can now be found using Ohm's Law.

$$E = I \times R$$

$$E = 24 \times .060$$

$$E = 1.44 \text{ volts}$$

This indicates that 1.44 volts will be used to push 24 amps through the resistance of the wire. If the air-conditioning unit is connected to 240 volts, the voltage at the unit will be 238.56 volts (240 − 1.44 = 238.56). This slight reduction in voltage will have no effect on the operation of the unit.

Now assume that the unit is located 100 feet from the panel, and that a #14 AWG wire was used. The resistance of the wire is:

$$R = \frac{K \times L}{CM}$$

$$R = \frac{10.4 \times 200}{4107}$$

$$R = .506 \text{ ohms}$$

The voltage drop of this wire will be:

$$E = I \times R$$

$$E = 24 \times .506$$

$$E = 12.144 \text{ volts}$$

The voltage appearing at the input of the unit is now 227.8 volts. This much reduction in voltage can cause a great deal of damage to the unit. This unit would probably have trouble during starting, since the current draw is greater at this time than any other.

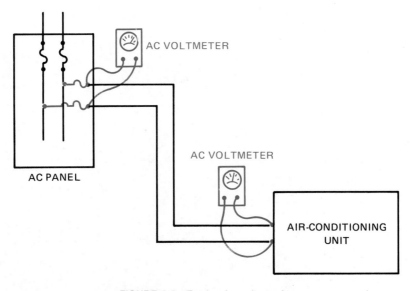

FIGURE 6-5 Testing for voltage drop

TESTING FOR EXCESSIVE VOLTAGE DROP

Testing a unit for excessive voltage can be done with a voltmeter. First, test the voltage at the panel with the unit turned off. Assume this voltage to be 240 volts. Next, start the air-conditioning unit and again check the voltage at the panel. If the voltage remains unchanged, it is an indication that there is no voltage drop at the panel and that all connections for that part of the circuit are good. If there is excessive voltage drop at the panel, it is an indication of bad connections, or the service entrance is too small for the load. For this example, assume the voltage remains at 240 volts when the unit is turned on. Next, check the voltage at the unit, figure 6-5. If there is a significant voltage drop at the unit, it indicates that the wire size is too small and that too much voltage is being used to push current through the wire. This problem can be corrected by connecting larger wires from the panel to the unit.

REVIEW QUESTIONS

1. Name four factors that determine the resistance of wire.

2. A wire has a diameter of .057 inches. What is its circular mil area?

3. What is a mil-foot of wire?

4. When the temperature of wire increases, does its resistance increase or decrease?

5. What determines the voltage rating of wire?

6. What two factors determine the amount of voltage rating a certain type of insulation will have?

7. How much resistance does 75 feet of #24 AWG wire have?

8. If a current of 4 amps flows through the wire in question #7, how much voltage will be dropped by the wire?

UNIT 7

Inductance

Alternating current circuits contain three basic types of loads. These are 1. resistive, 2. inductive, and 3. capacitive.

RESISTIVE CIRCUITS

The simplest of the AC loads is a circuit that contains only pure resistance, figure 7-1. In a pure-resistive circuit, the voltage and current are in phase with each other, figure 7-2. Voltage and current are in phase when they cross the zero line at the same point, and have their peak positive and negative values at the same time. A pure-resistive circuit is very similar to a direct-current circuit in the respect that true power or watts is equal to the voltage times the current. Examples of pure-resistive circuits are: the heating elements of an electric range,

an electric hot-water heater, and the resistive elements of an electric furnace.

INDUCTIVE CIRCUITS

An inductive circuit contains an inductor or coil as the load instead of a resistor, figure 7-3. The two most common types of inductive circuits are motors and transformers. Inductors are measured in units called the *henry*. The unit of in-

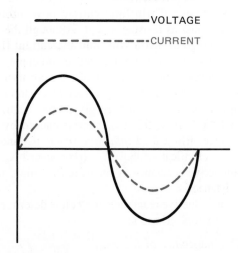

FIGURE 7-2 Current and voltage are in phase.

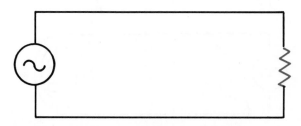

FIGURE 7-1 Pure resistive circuit

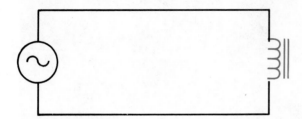

FIGURE 7-3 A pure inductive circuit

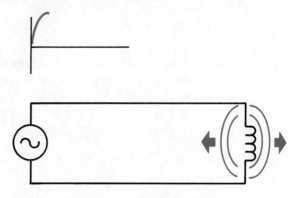

FIGURE 7-4 Magnetic field expands.

ductance is named in honor of Joseph Henry, a physicist who studied electricity. The electrical symbol for inductance is "*L*."

Inductors differ from resistors in several ways. One way is that the current of an inductor is not limited by the resistance of the coil. When an inductor is connected into an AC circuit and the voltage begins to rise from zero toward its peak value, a magnetic field is created around the coil, figure 7-4. As the expanding magnetic field cuts through the wires of the coil, a voltage is induced in the coil. *The voltage induced in a coil is always opposed to the voltage that creates it.* As the applied voltage begins to drop from its peak value back toward zero, the magnetic field around the inductor begins to collapse, figure 7-5. Notice that the induced voltage is opposite in polarity to the applied voltage. An induced voltage is 180° out of phase with the applied voltage, figure 7-6.

Since the induced voltage is opposed to the applied voltage, it will limit current flow through the circuit just as resistance will. Although the induced voltage of a coil will limit the current flow "like" resistance, it is not resistance and cannot be treated as resistance. The unit of measure used to describe the current-limiting effect of an induced voltage is *reactance* and is given the electrical symbol "*X*." Since this reactance is caused by an inductance, it is called *inductive reactance* and is given the electrical symbol "X_L" (pronounced X sub L). Inductive reactance is measured in ohms just as resistance is.

The inductive reactance of a coil is determined by two factors. These are:

1. The inductance of the coil,
2. The frequency of the applied voltage.

If these factors are known, a formula can be used to find the inductive reactance of the coil. This formula is:

$$X_L = 2 \times \pi \times F \times L$$

X_L = inductive reactance
π = the greek letter PI, which has a value of 3.1416
F = the frequency of the AC voltage
L = the inductance of the coil in henrys

EXAMPLE

In the circuit shown in figure 7-7, a coil has an inductance of .7 henrys, and is connected to a 120-volt, 60-Hz line. Find the current flow in the circuit.

To solve the problem, the first step is to find the amount of inductive reactance in the circuit.

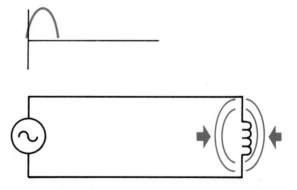

FIGURE 7-5 Magnetic field collapses.

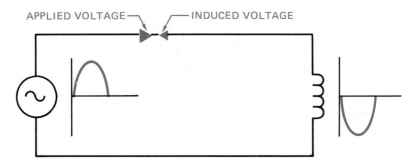

FIGURE 7-6 The induced voltage is opposed to the applied voltage.

$X_L = 2 \times \pi \times F \times L$
$X_L = 2 \times 3.1416 \times 60 \times .7$
$X_L = 263.9$ ohms

Now that the inductive reactance of the coil is known, the current flow in the circuit can be calculated. If the value of inductive reactance is used like resistance, the formula to find current in a pure-inductive circuit is $I = E/X_L$.

$$I = \frac{120}{263.9}$$

$I = .455$ amps

VOLTAGE AND CURRENT RELATIONS

As stated previously, the voltage and current in a pure-resistive circuit are in phase with each other. In a pure-inductive circuit, however, the current lags behind the voltage by 90°, figure 7-8. In this type of circuit there is no true power or watts. In a resistive circuit, the resistor limits the current

flow by converting the energy of the moving electrons into heat. This conversion of one form of energy into another represents a true power loss. In an inductive circuit, the energy of the moving electrons is stored in the magnetic field created around the inductor. When the magnetic field collapses, this energy is given back into the circuit. Notice that the resistor used the electrical energy by converting it into heat, but the inductor stored the energy and then returned it to the circuit.

IMPEDANCE

In an alternating-current circuit that contains only resistance, the current is limited by the value

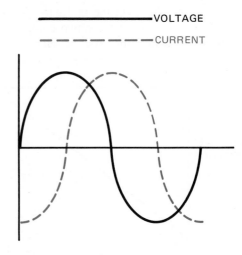

FIGURE 7-8 Current lags voltage by 90° in a pure inductive circuit.

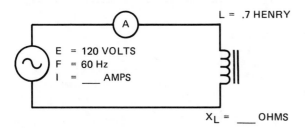

FIGURE 7-7 Current flow is limited by inductive reactance.

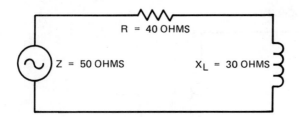

FIGURE 7-9 A series circuit containing resistance and inductance

of the resistor only. In this type of circuit the current flow can be calculated using the Ohm's Law formula $I = E/R$.

In an AC circuit that contains only inductance, the current is limited only by the value of inductive reactance. In this type of circuit, the current flow can be calculated using the formula $I = E/X_L$.

In a circuit like the one shown in figure 7-9, there are elements of both resistance and inductive reactance contained in the same circuit. In this type of circuit, it cannot be said that the current is limited by resistance, because there is also inductive reactance. It can also not be said that the current is limited by inductive reactance because of the resistance. Alternating-current circuits use a different value to represent the total amount of opposition to current flow in the circuit regardless of what type of components are found in the circuit. This value is known as *impedance* and is given the symbol "Z."

The resistor and inductor in figure 7-9 are connected in series. Since these two components are connected in series, they will be added. They

cannot be added in the normal way, however, because the inductance is not in phase with the resistance. Since inductive reactance is 90° out of phase with resistance, the total amount of opposition to current flow will be the value of the hypotenuse of the right triangle formed by the resistance and inductive reactance, figure 7-10. To find the total value of impedance in this circuit, the formula $Z = \sqrt{R^2 + X_L^2}$ can be used. If the resistor has a value of 40 ohms and the inductor has an inductive reactance of 30 ohms, the impedance will be:

$$Z = \sqrt{R^2 + X_L^2}$$
$$Z = \sqrt{40^2 + 30^2}$$
$$Z = \sqrt{1600 + 900}$$
$$Z = \sqrt{2500}$$
$$Z = 50 \text{ ohms}$$

APPARENT POWER

In a direct-current circuit, the true power or watts is always equal to the voltage multiplied by the current. This is because the current and voltage are never out of phase with each other. This also is true for an AC circuit that contains only pure resistance since the voltage and current are in phase.

In a circuit that contains pure inductance, however, there is no true power or watts. In this type of circuit the voltage multiplied by the current equals a value known as *vars*, which stands for (Volt-Amps Reactive). Vars is often referred to as wattless power.

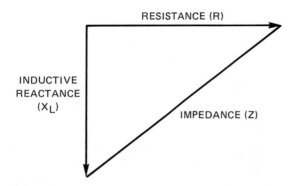

FIGURE 7-10 Impedance is the sum of resistance and inductive reactance.

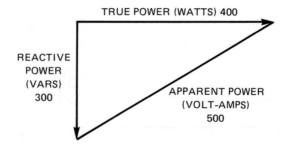

FIGURE 7-11 Volt-amps is the vector sum of watts and vars.

The apparent power or volt-amps of an AC circuit is the applied voltage multiplied by the current flow in the circuit. The amount of apparent power as compared to the true power or vars is determined by the elements of the circuit itself. In the circuit shown in figure 7-11, the amount of true power is 400 watts. The amount of reactive power is 300 vars. The apparent power is 500 volt-amps. Notice that the apparent power is found by adding the watts and vars together in the same manner that the resistance and inductive reactance were added to find the total value of impedance. Volt-amps can be calculated by the formula:

$$\text{Volt-amps} = \sqrt{W^2 + \text{Vars}^2}$$

POWER FACTOR

The power factor of an alternating-current circuit is a ratio of the apparent power compared to the true power. Power factor is important because utility companies charge industries large penalties for a poor power factor. The power factor of the circuit shown in figure 7-11 can be found by:

$$PF = \frac{W}{VA}$$

$$PF = \frac{400}{500}$$

$$PF = .8$$

$$PF = 80\%$$

Notice in this circuit, the power factor is 80%. This means 80% of the load is resistive and 20% is reactive. If the load is pure resistive, the power factor would be 100% or unity.

Utility companies become very concerned about power factor because they must furnish the amount of current needed to produce the volt-amp value. The company, however, is charged by the amount of true power or watts used. In this instance, if the applied voltage is 120 volts, the utility company must supply 4.16 amps to operate the load (500 volt-amps/120 volts = 4.16 amps). The actual amount of current being used to operate the load, however, is 3.33 amps (400 watts/120 volts = 3.33 amps). Since the air-conditioning load is often the major part of the electrical power consumed by an industry or office building, power factor can become an important consideration to the service technician.

REVIEW QUESTIONS

1. Name the three basic types of alternating-current loads.

2. What type of load always has its voltage and current in phase with each other?

3. In a pure-inductive circuit, how many degrees out of phase is the current with the voltage?

4. Does the current lead or lag the voltage in question #3?

5. What electrical value is used to measure inductance?

6. What is inductive reactance?

7. What electrical value is used to measure the total opposition to current flow in an AC circuit?

8. What is power factor?

UNIT 8
Capacitance

The third type of alternating-current load to be discussed is capacitance. A capacitor can be made by separating two metal plates with an insulating material, figure 8-1. The insulating material used to isolate the plates from each other is called the dielectric. There are three factors that determine how much capacitance a capacitor will have. These are:

1. The surface area of the plates,
2. The distance between the plates,
3. The type of dielectric material used between the plates.

CHARGING A CAPACITOR

In figure 8-2, the terminals of a capacitor have been connected to a battery. Electrons are negative particles. Therefore, the positive terminal of the battery attracts electrons from one plate of the capacitor. The negative terminal of the battery will cause electrons to flow to the other capacitor plate. This flow of current will continue until the voltage across the capacitor plates is equal to the battery voltage. If the battery is disconnected, the capacitor will be left in a charged state. CAUTION: IT IS THE HABIT OF SOME PEOPLE TO CHARGE A CAPACITOR TO A HIGH VOLTAGE AND THEN HAND THE CAPACITOR TO ANOTHER PERSON. WHILE SOME PEOPLE THINK THIS IS COMICAL, IT IS AN EXTREMELY DANGEROUS PRACTICE. CAPACITORS HAVE THE ABILITY TO SUPPLY AN ALMOST INFINITE AMOUNT OF CURRENT. UNDER CERTAIN CONDITIONS, A CAPACITOR CAN HAVE ENOUGH POWER TO CAUSE A PERSON'S HEART TO GO INTO FIBRILLATION.

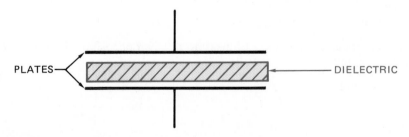

FIGURE 8-1 A capacitor is made with two metal plates separated by a dielectric.

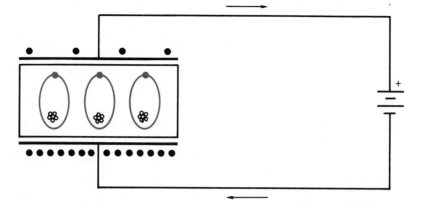

FIGURE 8-2 An electrostatic charge is stored in the atoms of the dielectric.

ELECTRO-STATIC CHARGE

Notice the illustration of the atoms in the dielectric material in figure 8-2. When a capacitor has been charged, the negative electrons of the dielectric material are repelled from the negative plate of the capacitor and attracted to the positive plate. This causes the electron orbit of the atoms in the dielectric to extend. This places the atoms of the dielectric material in tension. This is known as *dielectric stress*. Placing the atoms of the dielectric under stress has the same effect as drawing back a bow and arrow and holding it, figure 8-3.

The amount of dielectric stress is determined by the voltage between the plates. The greater the voltage, the greater the dielectric stress. If the voltage becomes too great, the dielectric will break down and destroy the capacitor. This is the reason capacitors have a voltage rating that must be followed.

The energy of a capacitor is stored in the dielectric and is known as an electro-static charge. It is this electrostatic charge that permits the capacitor to produce extremely high currents under certain conditions. If the leads of a charged capacitor are shorted together, it has the same effect as releasing the drawn bow in figure 8-3. The arrow will be propelled forward at great speed. The same is true for the electrons of the capacitor. When the electron orbits of the dielectric snap back, the electrons stored on the negative capacitor plate are propelled toward the positive plate at great speed.

CAPACITOR RATINGS

Capacitors are rated in units called the farad. The farad is actually such a large amount of capacitance it is not practical to use. For this reason a unit called the micro-farad is generally used. A micro-farad is one millionth of a farad. The Greek lowercase letter mu is used to symbolize micro, "μ." The term micro-farad is indicated by combining mu and lowercase "f," "μf." Since the letter mu is not included on a standard typewriter, the term micro-farad is sometimes shown as "uf" or "mf." All of these terms mean the same thing.

Another term used is the pica-farad. This term is used for extremely small capacitors found in

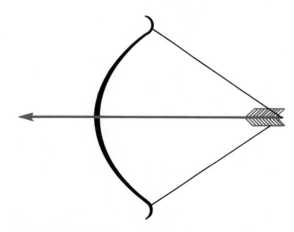

FIGURE 8-3 Dielectric stress is similar to drawing back a bow and arrow, and holding it.

FIGURE 8-4 A pure capacitive circuit

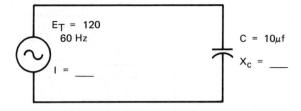

FIGURE 8-5 Capacitive reactance limits current flow.

electronics applications. A pica-farad is one millionth of a micro-farad, and is generally shown as "μμf" or "pf."

When AC voltage is applied to a capacitor, figure 8-4, the plates of the capacitor are alternately charged and discharged each time the current changes direction of flow. When a capacitor is charged, the voltage across its plates becomes the same as this applied voltage. As the voltage across the plates of a capacitor increases, it offers resistance to the flow of current. The applied voltage must continually overcome the voltage of the capacitor to produce current flow. The current in a pure-capacitive circuit is limited by the voltage of the charged capacitor. Since current is limited by a counter voltage and not resistance, the counter voltage of the capacitor is referred to as reactance. Recall that the symbol for reactance is "X." Since this reactance is caused by capacitance, it is called *capacitive reactance* and is symbolized by "X_c" (pronounced X sub c).

The amount of capacitive reactance in a circuit is determined by two factors. These are:

1. Frequency of the AC voltage,
2. The size of the capacitor.

If the frequency of the line and the capacitance rating of the capacitor are known, the capacitive reactance can be found using the following formula:

$$X_c = \frac{1}{2 \times \pi \times F \times C}$$

The value of capacitive reactance is measured in ohms. In the formula to find capacitive reactance:

X_c = Capacitive Reactance

π = The Greek letter PI, which has a value of 3.1416

F = Frequency in Hz

C = The value of capacitance in farads. Since most capacitors are rated in micro-farads, be sure to write the capacitance value in farads. This can be done by dividing the micro-farad rating by 1,000,000, or moving the decimal point six places to the left. Example: to change a 50 μf capacitor to a value expressed in farads, move the decimal point after the 50, six places to the left. This capacitor has a value of .000050 farads.

EXAMPLE

Find the current flow in the circuit shown in figure 8-5.

Solution: To find the current flowing in this circuit, the amount of capacitive reactance of the capacitor must first be found.

$$X_c = \frac{1}{2 \times \pi \times F \times C}$$

$$X_c = \frac{1}{2 \times 3.1416 \times 60 \times .000010}$$

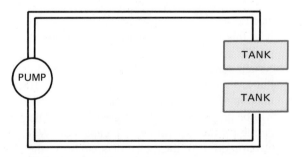

FIGURE 8-6 Water flows in this system in a manner similar to the way current flows in a capacitive circuit.

$$X_c = \frac{1}{.0037699}$$

$$X_c = 265.2 \text{ ohms}$$

Now that the capacitive reactance of the circuit is known, the value of current can be found using the formula: $I = E/X_c$.

$$I = \frac{120}{265.2}$$

$$I = .452 \text{ amps}$$

CURRENT FLOW IN A CAPACITIVE CIRCUIT

Notice that a capacitor is constructed of two metal plates separated by an insulator. One of the metal plates is connected to one side of the circuit, and the other metal plate is connected to the other side of the circuit. Since there is an insulator separating the two plates, current cannot flow through a capacitor. When a capacitor is connected into a direct-current circuit, current will flow until the capacitor has been charged to the value of the applied voltage, and then stop. When a capacitor is connected into an alternating-current circuit, current will "appear" to flow through the capacitor. This

is because the plates of the capacitor are alternately charged and discharged each time the current reverses direction. To understand this concept better, refer to the water circuit shown in figure 8-6. In this illustration, a water pump is connected to two tanks. The pump is used to pump water back and forth between the two tanks. When one tank becomes full, the direction of the pump is reversed and water is pumped from the full tank back into the empty tank. Notice that there is no complete loop in this hydraulic circuit for water to flow from one side of the pump to the other, but water does flow because it is continuously pumped from one tank to the other.

CURRENT AND VOLTAGE RELATIONSHIPS

In a pure-capacitive circuit, the voltage and current are out of phase with each other. Figure 8-7 shows that the current in a pure-capacitive circuit leads the voltage by 90°. Since the voltage and current are 90° out of phase with each other, there is no true power or watts consumed in a pure-capacitive circuit. The capacitor stores the energy in an electro-static field, and then returns it to the circuit at the end of each half cycle.

In the circuit shown in figure 8-8, a resistor and capacitor are connected in series with each other. Since this circuit contains elements of both resistance and capacitive reactance, the current is limited by impedance. The impedance for a circuit of this type can be found by using the formula $Z = \sqrt{R^2 + X_c^2}$. Notice this is the same basic formula

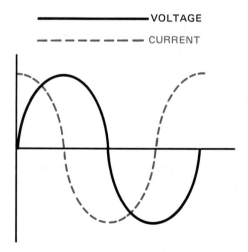

VOLTAGE

---- CURRENT

FIGURE 8-7 Current leads voltage by 90° in a pure capacitive circuit.

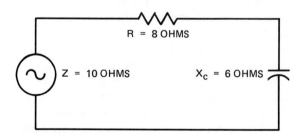

R = 8 OHMS

Z = 10 OHMS

X_c = 6 OHMS

FIGURE 8-8 Impedance must be used to determine the current flow in a circuit that contains resistance and capacitive reactance.

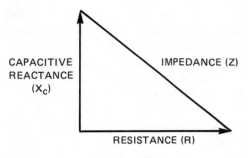

CAPACITIVE
REACTANCE
(X_c)

IMPEDANCE (Z)

RESISTANCE (R)

FIGURE 8-9

as the one used to find the impedance of a circuit that contains both resistance and inductive reactance. The impedance of the circuit shown in figure 8-8 can be found by the following:

$$Z = \sqrt{R^2 + X_c^2}$$
$$Z = \sqrt{8^2 + 6^2}$$
$$Z = \sqrt{64 + 36}$$
$$Z = \sqrt{100}$$
$$Z = 10 \text{ ohms}$$

Figure 8-9 shows a vector diagram of the circuit in figure 8-8.

POWER FACTOR CORRECTION

Since the current flow in a capacitive circuit leads the voltage by 90° and the current in an inductive circuit lags the voltage by 90°, the current of a capacitive circuit is in direct opposition to the current of an inductive circuit. Figure 8-10 illustrates the currents of capacitive and inductive circuits as compared to each other. These two currents are 180° out of phase with each other. When the capacitive current is at its peak positive value, the inductive current is at its peak negative value. When the capacitive current is at its peak negative value, the inductive current is at its peak positive value. Since these two currents are in direct opposition, one can be used to cancel the other.

The circuit shown in figure 8-11 shows a parallel circuit that contains a resistor, an inductor, and a capacitor. The applied voltage of the circuit is 120 volts at 60 Hz. Since this is a parallel cir-

cuit, the voltage applied to each component will be the same—120 volts. The resistor has a resistance of 12 ohms. This permits a current flow of 10 amps through the resistor ($120/12 = 10$). The inductor has an inductive reactance of 24 ohms. This permits a current flow through the inductor of 5 amps ($120/24 = 5$). The capacitor has a capacitive reactance of 24 ohms. This permits a current flow through the capacitor of 5 amps.

QUESTION

What is the total current flow in the circuit? In a parallel circuit, current is added. Therefore, it would appear that the current flow would be 20 amps ($10 + 5 + 5 = 20$). The currents of this circuit, however, are out of phase with each other. Figure 8-12 shows a vector diagram of this circuit. Notice that the 5 amps of capacitive current is 180° out of phase with the 5 amps of inductive current. These two currents will cancel each other. The AC alternator sees only the resistance in this circuit.

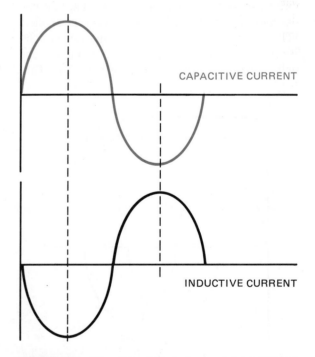

CAPACITIVE CURRENT

INDUCTIVE CURRENT

FIGURE 8-10 Capacitive and inductive current are 180° out of phase with each other.

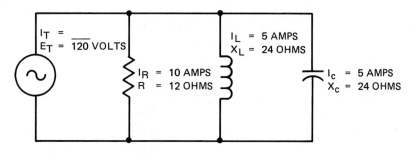

FIGURE 8-11 A parallel circuit has resistance, inductance, and capacitance.

The current is, therefore, the same as the current flow through the resistor, or 10 amps.

POWER FACTOR CORRECTION OF A MOTOR

In the circuit shown in figure 8-13, an AC induction motor is connected to a 120-volt line. A wattmeter is used to measure the amount of true power in the circuit. For this example it will be assumed that the wattmeter has a reading of 720 watts. An ammeter has also been inserted in the circuit. Assume the ammeter has a reading of 10 amps. The apparent power or volt-amp value for this circuit is 1200 VA (120 volts × 10 amps = 1200 VA). The power factor of this circuit can now be computed using the formula ($PF = W/VA$).

$$PF = W/VA$$

$$PF = 720/1200$$

$$PF = .6 \text{ or } 60\%$$

If the power factor of this motor is to be corrected, it must be determined how much of this circuit is comprised of true power and how much is comprised of reactive power. Since the true power (watts) and the apparent power (volt-amps) is known, the reactive power (vars) can be found using the following formula:

$$Vars = \sqrt{VA^2 - W^2}$$

$$Vars = \sqrt{1200^2 - 720^2}$$

$$Vars = \sqrt{1,440,000 - 518,400}$$

$$Vars = \sqrt{921,600}$$

$$Vars = 960$$

Since a motor is an inductive device, the reactive power in this circuit can be canceled by an equal amount of capacitive vars. If a capacitor of the correct value is connected in parallel with the motor, the power factor will be corrected. To find the correct value of capacitance, determine the amount of capacitance needed to produce a var reading of 960. The amount of capacitive reactance can be found using the formula:

CAPACITIVE CURRENT 5 AMPS

RESISTIVE CURRENT
10 AMPS

INDUCTIVE CURRENT 5 AMPS

FIGURE 8-12 Capacitive current and inductive current are 180° out of phase with each other.

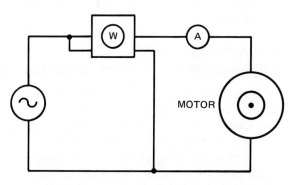

FIGURE 8-13 Finding power factor of a motor

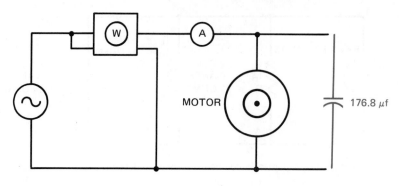

FIGURE 8-14 A capacitor corrects motor power factor.

$$X_c = E^2/\text{vars}$$

$$X_c = 120^2/960$$

$$X_c = 15 \text{ ohms}$$

The amount of capacitance needed to produce 15 ohms of capacitive reactance at 60 Hz can be calculated using the following formula:

$$C = \frac{1}{2 \times \pi \times F \times X_c}$$

$$C = \frac{1}{2 \times 3.1416 \times 60 \times 15}$$

$$C = \frac{1}{5654.88}$$

$$C = .0001768 \text{ farads}$$

The answer for the value of C is in farads. To convert farads to micro-farads, multiply the answer by 1,000,000, or move the decimal point 6 places to the right. .0001768 farads becomes 176.8 μf. If a capacitor of this value is connected in parallel with the motor as shown in figure 8-14, the power factor will be corrected.

CAPACITOR TYPES

The most common types of capacitors used in the air-conditioning field fall into two categories. One kind is known as an oil-filled type. Figure 8-15 shows a photograph of this type of capacitor. The oil-filled capacitor is made with two metal foil plates separated by paper. The paper is soaked in a special dielectric oil. These capacitors are true

AC capacitors and are generally used as the run capacitors on many single-phase air-conditioning compressors. They are also used as the starting capacitors on some units. The important ratings on these capacitors are the micro-farad rating and the voltage rating. The voltage rating of a capacitor should never be exceeded. It is permissible to use a capacitor of higher voltage rating, but never use a capacitor with less voltage rating.

The second type of capacitor frequently used in air-conditioning systems is the AC electrolytic capacitor. The AC electrolytic capacitor is used as the starting capacitor on many small single-phase motors. This type of capacitor is designed to be used for a short period of time only. If an AC electrolytic capacitor were to be used in a continuous circuit, such as the running capacitor of a com-

FIGURE 8-15 Oil-filled capacitor (Courtesy Westinghouse Electric Corp.)

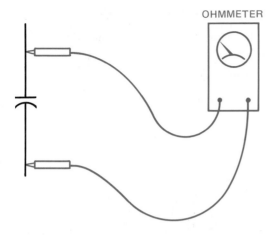

FIGURE 8-16 Testing a capacitor with an ohmmeter

FIGURE 8-17 Meter for testing capacitance and inductance (Courtesy of Sencore)

pressor, it would fail in a short period of time. The advantage of the AC electrolytic capacitor is that a large amount of capacitance can be housed in a small case size. This makes the AC electrolytic capacitor a good choice for starting circuits, because the capacitor is in the circuit for only a few seconds when the motor is started.

TESTING A CAPACITOR

Capacitors can be tested for a short with an ohmmeter. If an ohmmeter is connected across the terminals of a capacitor as shown in figure 8-16, the meter should show a deflection up scale and then return to infinity ohms. The deflection up scale indicates current flow to the capacitor when it is being charged by the ohmmeter battery. If the leads of the ohmmeter are reversed, the meter should deflect twice as far up scale and then return to infinity ohms.

The ohmmeter test for a capacitor is actually a very poor test. This test will not indicate a broken plate that would lower the value of capacitance. The ohmmeter test will not indicate if the dielectric of the capacitor is being broken down under voltage. To test a capacitor, a testing instrument specifically designed for this job should be used. A capacitor tester is shown in figure 8-17.

REVIEW QUESTIONS

1. What three factors determine the capacitance of a capacitor?

2. What is the dielectric?

3. In what type of field is the energy of a capacitor stored?

4. In a pure-capacitive circuit, how many degrees are the current and voltage out of phase with each other?

5. Does a capacitive current lead the voltage or lag the voltage?

6. What limits the current in a capacitive circuit?

7. Name two common types of capacitors used in the air-conditioning field.

8. What type of capacitor is generally used as the running capacitor on many air-conditioning compressors?

9. What is the advantage of an AC electrolytic capacitor?

10. What is the disadvantage of an AC electrolytic capacitor?

SECTION 2

Single-Phase Motors

UNIT 9
Split-Phase Motors

Since three-phase power is not available to small business and residential locations, the air-conditioning equipment for these areas is powered by single-phase electric motors. The single-phase motors used in air-conditioning systems are generally one of two types. These are the split-phase and the shaded-pole induction motor.

SPLIT-PHASE MOTORS

Split-phase motors fall into three general classifications. These are:

1. The resistance-start induction run motor,
2. The capacitor-start induction run motor,
3. The permanent-split capacitor motor (PSC).

Although all of these motors have different operating characteristics, they are similar in construction. Split-phase motors get their name from the manner in which they operate. Recall from the study of three-phase motors that the basic principle of operation of an AC induction motor is that of a rotating magnetic field. One of the factors that causes the field to rotate is the fact that the three-line voltages are 120° out of phase with each other. In a single-phase power system, there is no other phase that can be used to produce a rotating field.

THE TWO-PHASE SYSTEM

In some parts of the world, two-phase power is produced. A two-phase system is produced by having an alternator with two sets of coils wound 90° out of phase with each other, figure 9-1. The voltages of a two-phase system are, therefore, 90° out of phase with each other, figure 9-2. The two out-of-phase voltages can be used to produce a rotating magnetic field. Since there have to be two voltages or currents out of phase with each other

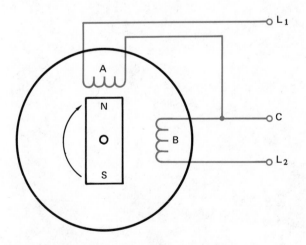

FIGURE 9-1 Two-phase alternator

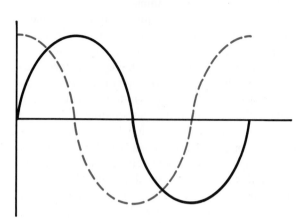

FIGURE 9-2 Two-phase voltages are 90° out of phase with each other.

FIGURE 9-3 Stator windings used in single-phase motors (Courtesy of Bodine Electric Co.)

to produce a rotating magnetic field, single-phase motors use two separate windings and create a phase difference between the currents in each of these windings. These motors literally "split" one phase and produce a second phase, hence the name split-phase motor.

STATOR WINDINGS

The stator of a split-phase motor contains two separate windings, the start winding and the run winding. The start winding is made of small wire and is placed near the top of the stator core. The run winding is made of relatively large wire and is placed in the bottom of the stator core. Figure 9-3 shows a photograph of a single-phase stator. Notice the difference in size and position of the two windings.

The fact that the start winding is made from small wire and placed near the top of the stator core causes it to have a higher resistance than the run winding. Notice that these two windings are connected in parallel with each other, figure 9-4. Since the run winding is made from large wire and placed near the bottom of the stator core, it is more inductive than the start winding. This causes the current flowing through the run winding to be out

of phase with the current in the start winding. These two out-of-phase currents produce the rotating magnetic field in the stator. The speed of the rotating magnetic field is determined by the same two factors that determined the synchronous field speed in a three-phase motor—the number of stator poles and the frequency.

Once the squirrel cage rotor, shown in figure 9-5, has accelerated to a point that is operating at about 75% of the rotating field speed, the start winding can be disconnected from the circuit. The rotor will continue to rotate due to the changing magnetic polarities of the run winding.

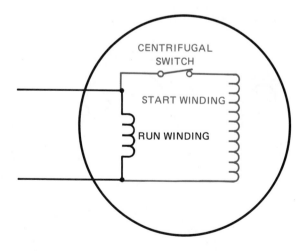

FIGURE 9-4 The start winding is connected in parallel with the run winding.

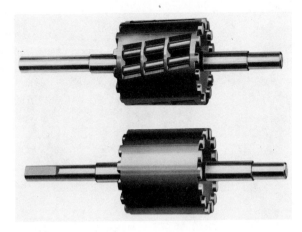

FIGURE 9-5 Squirrel cage rotor (Courtesy of Bodine Electric Co.)

TESTING THE STATOR WINDING

The stator winding of a single-phase motor is generally tested with an ohmmeter. The ohmmeter test can be used to determine if a winding is open or grounded. Many single-phase motors have one lead of the run and start windings connected as shown in figure 9-6. To test the windings for an open, connect one ohmmeter lead to the common motor terminal, and the other meter lead to the run winding. The ohmmeter should indicate continuity through the winding. The resistance of the run winding of a single-phase motor can vary greatly from one motor to another. The winding resistance of a single-speed motor may be only one or two ohms, while the resistance of a multi-speed fan motor may be 10 to 15 ohms.

To test the start winding for an open, connect the ohmmeter leads to the common terminal and the S terminal. The start winding should indicate continuity, and should have a higher resistance than the run winding. This difference of resistance may not be great, but the start winding should have a higher resistance than the run winding.

To test the stator winding for a ground, connect one of the ohmmeter leads to the case of the motor, figure 9-7. Alternately check each motor terminal with the other ohmmeter lead. The ohmmeter should indicate no continuity between either winding and the case of the motor.

A shorted start winding can sometimes be detected by the fact that the motor will not start, but will run if the shaft is turned by hand. The motor will produce a humming sound but will not turn when power is first applied to it. The shaft can be turned in either direction by hand and the motor will continue to run in that direction.

REVERSING DIRECTION OF ROTATION

The direction of rotation of a split-phase motor can be reversed by changing the start winding leads or the run winding leads, but not both. The rotation is generally reversed by changing the start

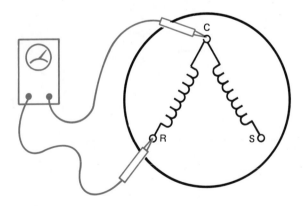

FIGURE 9-6 Testing the split-phase motor for an open winding.

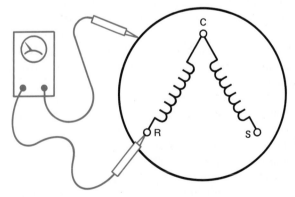

FIGURE 9-7 Testing a split-phase motor for a grounded winding

winding leads with respect to the run winding. Some motor manufacturers bring both start winding leads to the outside of the motor. This permits the service technician to decide the direction of rotation the motor is to turn when it is installed on the unit.

DISCONNECTING THE START WINDINGS

As stated previously, when the rotor of a split-phase motor reaches about 75% of the speed of the rotating magnetic field, the start windings can be disconnected from the circuit. Open-case motors generally use a centrifugal switch to perform this job. A simple centrifugal switch mechanism is shown in figure 9-8. This diagram is intended to illustrate the principal of operation of the switch. When the shaft is not turning, the bottom ring rides against the movable contact arm. The weight of the metal balls overcomes the spring tension pushing upward against the movable contact arm, and the contact is held closed. When the shaft turns, centrifugal force causes the metal balls to spin outward. As the metal balls spin outward, the bottom

ring is mechanically lifted away from the movable contact. This permits the spring to open the contact.

Figure 9-9 shows a photograph of a centrifugal switch used on a split-phase motor. This switch must overcome the force of a spring, instead of gravity, to move the disk from one position to another. When the counter weights are turned fast enough, the spring causes the disk to snap away from the contact and instantly open the switch to the start windings. When the motor is turned off, the disk will snap back into position when the motor speed decreases enough to permit the force of

FIGURE 9-9A & B Centrifugal switch (Courtesy of Westinghouse Electric Corp.)

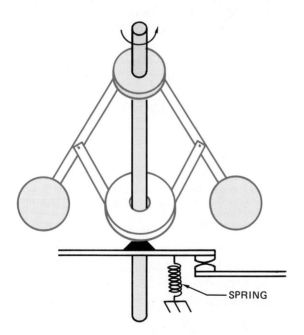

FIGURE 9-8 Basic centrifugal switch

the spring to overcome the centrifugal force of the counterweights.

Hermetically sealed motors, such as those used in hermetically sealed compressors, cannot use a centrifugal switch to disconnect the start windings. Starting relays, which will be discussed later, are used to disconnect the start windings on these types of motors.

DUAL-VOLTAGE MOTORS

Single-phase motors can also be constructed to operate on two separate voltages. These motors are designed to be connected to 120 or 240 volts. A common connection for this type of motor contains two run windings and one start winding, figure 9-10. The run windings are labeled T1–T2, and T3–T4. The start winding is labeled T5 and T6. In the circuit shown in figure 9-10, the windings have been connected for operation on a 240-volt line. Each winding is rated at 120 volts. The two run windings are connected in series, which causes each to have a voltage drop of 120 volts when connected to 240 volts. Notice that the start winding has been connected in parallel with one of the run windings. This causes the start winding to have an applied voltage of 120 volts also. Notice that each of the windings has 120 volts connected to it, which is the rating of the windings.

If the motor is to be operated on a 120-volt line, the windings are connected in parallel as shown

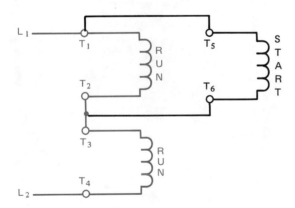

FIGURE 9-10 The run windings are connected in series for a high-voltage connection.

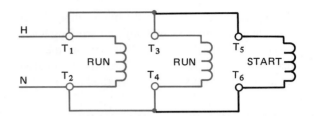

FIGURE 9-11 All windings are connected in parallel for a low-voltage connection.

in figure 9-11. Since these windings are connected in parallel, each will have 120 volts applied to it.

MOTOR POWER CONSUMPTION

It should be noted that the motor does not use less energy when connected to 240 volts than it does when connected to 120 volts. Power is measured in watts, and the watts will be the same regardless of the connection. When the motor is connected to operate on 240 volts, it will have half the current draw as it does on a 120-volt connection. Therefore, the amount of power used is the same. For example, assume the motor has a current draw of 5 amps when connected to 240 volts and 10 amps when connected to 120 volts. Watts can be computed from multiplying volts by amps. When the motor is connected to 240 volts, the amount of power used is 240 × 5 = 1200 watts. When the motor is connected to 120 volts, the power used is 120 × 10 = 1200 watts. The 240-volt connection is generally preferred, however, because the lower current draw causes less voltage drop on the line supplying power to the motor. If the motor is located a long distance from the panel, voltage drop of the wire can become very important to the operation of the unit.

RESISTANCE START INDUCTION-RUN MOTOR

The resistance start induction-run motor develops a rotating magnetic field by using the resistance of the start winding. Since the start winding has a higher resistance than the run winding, the current in the start winding will be out of phase with the current in the run winding. Maximum starting torque is developed when these two currents are 90° out of phase with each other. The run winding, however, is not a pure inductor. It has some resistance in the wire used to make it. This causes the current flowing in the run winding to be less than 90° out of phase with the voltage.

The start winding is not a pure resistance. This winding has some inductance, which causes it to be out of phase with the voltage. As a result, there is not a 90° phase difference between the current in the start winding and the current in the run winding, figure 9-12. The actual phase angle difference between these two currents in this type of motor is about 40°. This causes the resistance start motor to have a relatively poor starting torque.

CAPACITOR START INDUCTION-RUN MOTOR

The capacitor start induction-run motor uses a capacitor to improve the starting torque. The capacitor is connected in series with the start winding, figure 9-13. Since this capacitor is in the circuit for only the amount of time needed to start the motor, an AC electrolytic capacitor is generally used as the starting capacitor for this motor. This limits the times this motor should be started over a short

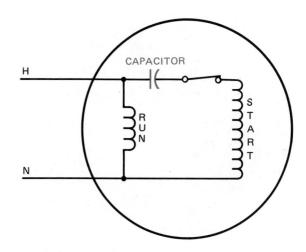

FIGURE 9-13 A schematic for a capacitor start induction-run motor

period of time. Most manufacturers do not recommend starting this type of motor more than eight times per hour, or the starting capacitor could be damaged.

The capacitor causes an increase in starting torque by shifting the phase angle of the current flowing in the start winding. The start winding circuit is now capacitive, which causes the current to lead the voltage. This produces a greater phase-angle difference between the run and start winding currents, figure 9-14. If the correct amount of capacitance is used, a phase-angle difference of 90° can be developed between these two currents, which results in maximum starting torque.

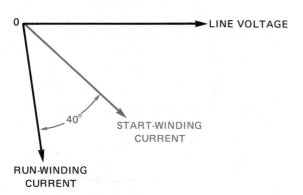

FIGURE 9-12 Start-winding and run-winding currents are about 40° out of phase with each other.

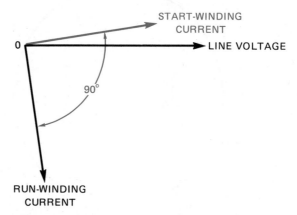

FIGURE 9-14 Start-winding and run-winding currents are about 90° out of phase with each other.

FIGURE 9-15 Capacitor start induction-run motor

FIGURE 9-17 Permanent-split capacitor motor

If the amount of capacitance is too great, the start winding current will become greater than 90° out of phase with the run winding current, and the starting torque will decrease. When replacing the starting capacitor for this type of motor, the microfarad rating recommended by the manufacturer should be used. If it is necessary, a capacitor of higher voltage rating can be used, but a capacitor of lower voltage rating should never be used. Figure 9-15 shows a photograph of a capacitor start induction-run motor.

PERMANENT-SPLIT CAPACITOR MOTOR

The permanent-split capacitor motor has greatly increased in popularity for use in the air-conditioning field over the past years. This type of split-phase motor does not disconnect the start windings from the circuit when it is running. This eliminates the need for a centrifugal switch or starting relay to disconnect the start windings from the circuit when the motor reaches about 75% of its full speed,

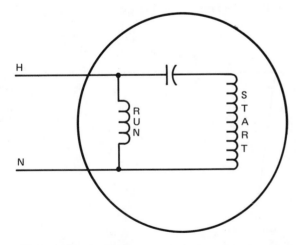

FIGURE 9-16 A schematic for a permanent-split capacitor motor

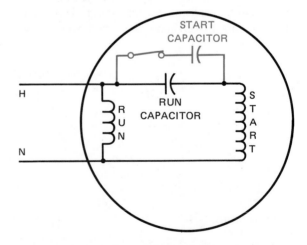

FIGURE 9-18 An extra starting capacitor used with a permanent-split capacitor motor

SECTION 2 SINGLE-PHASE MOTORS

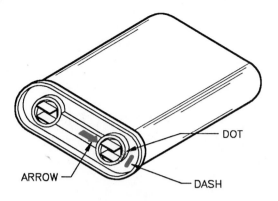

FIGURE 9-19 Markings indicate plate nearest capacitor case.

figure 9-16. This motor has good starting torque and good running torque. Since the capacitor remains in the circuit during operation, it helps correct power factor for the motor.

The capacitor used in this type of motor is generally the AC oil-filled type. A photograph of a permanent-split capacitor motor is shown in figure 9-17. Since this capacitor remains connected in the circuit, an AC electrolytic capacitor cannot be used to replace the run capacitor for this type of motor.

The permanent-split capacitor motor will sometimes use an extra capacitor to aid in starting. When this is done, the start capacitor is connected in parallel with the run capacitor. During the time of starting, both of these capacitors are connected

in the circuit, figure 9-18. When the motor has accelerated to about 75% of full speed, the start capacitor is disconnected from the circuit. If the motor is an open type, the start capacitor will be disconnected by a centrifugal switch. If the motor is sealed, such as a hermetically-sealed compressor, the start capacitor will be disconnected by a starting relay.

IDENTIFYING CAPACITOR TERMINALS

Most run capacitors and some starting capacitors are of the oil-filled type, figure 8-15. This is especially true for high current motors such as those used to operate compressors. Many manufacturers of oil-filled capacitors will identify one terminal with an arrow, a painted dot, or by stamping a dash in the capacitor can, figure 9-19. This identified terminal marks the connection to the plate that is located nearer to the metal container or can. It has long been known that when a capacitor's dielectric breaks down and permits a short circuit to ground, it is most often the plate nearer to the outside case that becomes grounded. For this reason, it is desirable to connect the identified capacitor terminal to the line side instead of to the motor start winding.

In figure 9-20, the run capacitor has been connected in such a manner that the identified terminal is connected to the start winding of a compressor motor. If the capacitor shorts to ground, a current

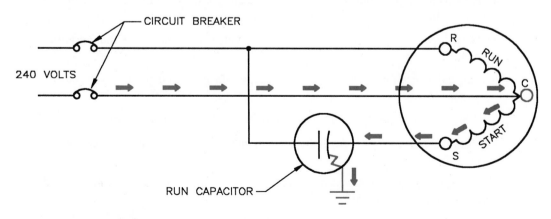

FIGURE 9-20 Identified capacitor terminal connected to motor start winding

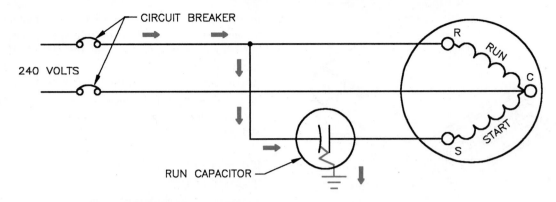

FIGURE 9-21 Identified capacitor terminal connected to the line

path would exist through the motor start winding. The start winding is an inductive-type load and inductive reactance will limit the value of current flow to ground. Since the flow of current is limited, it will take the circuit breaker or fuse time to open the circuit and disconnect the motor from the power line. This time delay can permit the start winding to overheat and become damaged.

In figure 9-21, the run capacitor has been connected in such a manner that the identified terminal is connected to the line side. If the capacitor shorts to ground, a current path would exist directly to ground, bypassing the motor start winding. When the capacitor is connected in this manner, the start winding does not limit current flow and allows the fuse or circuit breaker to open almost immediately.

REVIEW QUESTIONS

1. What is a split-phase motor?

2. What are the three basic types of split-phase motors?

3. Explain the difference in construction of run windings and start windings.

4. How many degrees out of phase should be current in the start winding with the current in the run winding to develop maximum starting torque?

5. What type of capacitor is generally used with a capacitor start induction-run motor?

6. Can the micro-farad value of this capacitor be increased to improve starting torque?

7. What type of capacitor is used with a permanent-split capacitor motor?

8. Does the capacitor of a capacitor start induction-run motor help correct power factor?

9. If necessary, can an AC electrolytic capacitor of higher voltage rating be used as the starting capacitor?

10. What is a centrifugal switch used for?

UNIT 10

The Shaded-Pole Induction Motor

The shaded-pole induction motor is another type of AC single-phase motor used to a large extent in the air-conditioning field. This motor is popular because of its simplicity and long life. The shaded-pole motor contains no start winding or centrifugal switch. The rotating magnetic field is created by a *shading coil* wound around one side of each pole piece.

THE SHADING COIL

The shading coil is wound around one end of the pole piece, figure 10-1. The shading coil is actually a large loop of copper wire or a copper band. Both ends of the loop are connected together to form a complete circuit. The shading coil acts in the same manner as a transformer with a shorted secondary winding. When the voltage of the AC waveform increases from zero toward its positive peak, a magnetic field is created in the pole piece. As magnetic lines of flux cut through the shading coil, a voltage is induced in the coil. Since the coil is a low-resistance short circuit, a large amount of current flows in the loop. This current flow causes an opposition to the change of magnetic flux, figure 10-2. As long as voltage is induced into the

shading coil, there will be an opposition to the change of magnetic flux.

When the AC voltage reaches its peak value, it is no longer changing and there is no voltage being induced into the shaded coil. Since there is no current flow in the shading coil, there is no opposition to the magnetic flux. The magnetic flux of the pole piece is now uniform across the pole face, figure 10-3.

When the AC voltage begins to decrease from its peak value back toward zero, the magnetic field of the pole piece begins to collapse. A current is again induced into the shading coil. The induced

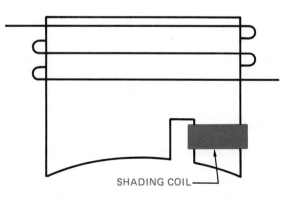

SHADING COIL

FIGURE 10-1 A shaded pole

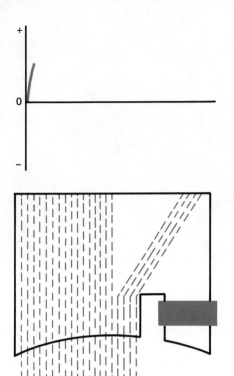

FIGURE 10-2 The shading coil opposes a change of magnetic flux as voltage increases.

current opposes the change of magnetic flux, figure 10-4. This causes the magnetic flux to be concentrated in the shaded section of the pole piece.

When the AC voltage passes through zero and begins to increase in the negative direction, the same set of events happen, except that the polarity of the magnetic field is reversed. If these events were to be seen in rapid order, it could be seen that the magnetic field rotates across the face of the pole piece.

SPEED

The speed of the shaded-pole induction motor is determined by the same factors that determine the synchronous speed of other induction motors;

frequency and number of stator poles. Shaded-pole motors are commonly wound as four- or six-pole motors. Figure 10-5 shows a drawing of a four-pole motor.

REVERSING DIRECTION OF ROTATION

The direction the magnetic field moves across the face of the pole piece is determined by the side of the pole piece that has the shaded coil. The rotor will turn in the direction of the shaded pole as shown by the arrow in figure 10-5. If the direction of rotation must be changed, ti can be done by remov-

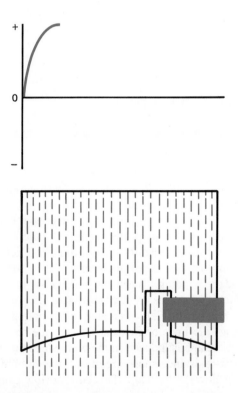

FIGURE 10-3 There is no opposition to magnetic flux when the voltage is not changing.

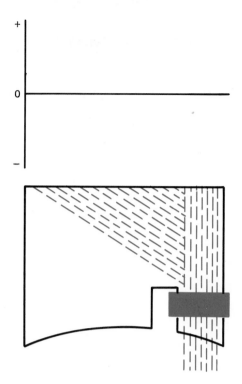

FIGURE 10-4 The shading coil opposes a change of flux when the voltage decreases.

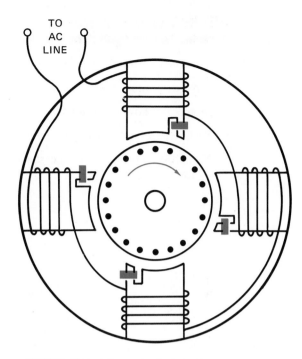

FIGURE 10-5 A four-pole shaded-pole induction motor

ing the stator winding and turning it around. This is not a common practice, however. As a general rule, the shaded-pole induction motor is considered to be nonreversible.

GENERAL OPERATING CHARACTERISTICS

The shaded-pole motor contains a standard squirrel cage rotor. The amount of torque produced is determined by the strength of the magnetic field of the stator, the strength of the magnetic field of the rotor, and the phase-angle difference between rotor current and stator current. The shaded-pole motor has a low starting torque and running torque. This motor is generally used in applications that do not require a large amount of starting torque, such as fans and blowers. Figure 10-6 shows a photograph of a shaded-pole induction motor.

FIGURE 10-6 Stator winding and rotor of a shaded-pole induction motor (Courtesy of Westinghouse Electric Corp.)

REVIEW QUESTIONS

1. What is a shading coil?

2. What determines the synchronous speed of a shaded-pole motor?

3. In general, how is the direction of a shaded-pole induction motor reversed?

4. What type of rotor does the shaded-pole motor contain?

5. Name two advantages of the shaded-pole motor over the split-phase induction motor.

UNIT 11

Multi-Speed Motors

Multi-speed AC motors have been used to a great extent in the air-conditioning field for many years. There are two basic types of multi-speed motors used. One type is known as the consequent pole motor. The other type is generally a permanent-split capacitor motor.

THE CONSEQUENT POLE MOTOR

The speed of the rotating magnetic field of an AC induction motor can be changed in either of two ways. These are:

1. Change the frequency of the AC voltage,
2. Change the number of stator poles.

The consequent pole motor changes the motor speed by changing the number of its stator poles. The run winding in figure 11-1 has been tapped in the center. If the AC line is connected to each end of the winding as shown, current flows through the winding in only one direction. Therefore, only one magnetic polarity is produced in the winding. If the winding is connected as shown in figure 11-2, current flows in opposite directions in each half of the winding. Since current flows through each half of the winding in opposite directions, the polarity of the magnetic field is different in each half of the winding. The run winding now has two polarities instead of one. There are now two magnetic poles

instead of one. If the windings of a two-pole motor were to be tapped in this manner, the motor could become a four-pole motor. The synchronous speed of a two-pole motor is 3600 RPM, and the synchronous speed of a four-pole motor is 1800 RPM.

The consequent pole motor has the disadvantage of having a wide variation in speed. When the speed is changed, it changes from a synchronous speed of 3600 RPM to 1800 RPM. The speed cannot be changed by a small amount. This wide variation in speed makes the consequent pole motor unsuitable for some loads, such as fans and blowers.

The consequent pole motor, however, does have some advantages over the other type of multi-

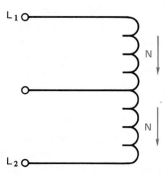

FIGURE 11-1 Center-tapped run winding

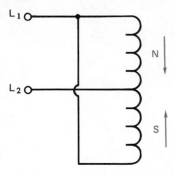

FIGURE 11-2 Two magnetic poles are produced.

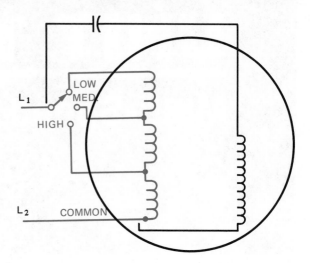

FIGURE 11-3 Three-speed fan motor

speed motor. When the speed of the consequent pole motor is reduced, its torque increases. For this reason, the consequent pole motor can be used to operate heavy loads, such as two-speed compressors.

MULTI-SPEED FAN MOTORS

Multi-speed fan motors have been used in the air-conditioning industry for many years. These motors are generally wound for two to five steps of speed, and are used to operate fans and squirrel cage blowers. A schematic drawing of a three-speed motor is shown in figure 11-3. Notice that the run winding has been tapped to produce low, medium, and high speed. The start winding is connected in parallel with the run winding section. The other end of the start lead is connected to an external oil-filled capacitor. This motor obtains a change in speed by inserting inductance in series with the run winding. The actual run winding for this motor is between the terminals marked High and C. The windings shown between High and Medium are connected in series with the main run winding. When the rotary switch is connected to the medium-speed position, the inductive reactance of this coil limits the amount of current flow through the run winding. When the current of the run winding is reduced, the strength of the magnetic field

of the run winding is reduced and the motor produces less torque. This causes the motor speed to decrease.

If the rotary switch is changed to the low position, more inductance is connected in series with the run winding. This causes less current to flow through the winding and another reduction in torque. When the torque is reduced, the motor speed decreases again.

Common speeds for a four-pole motor of this type are 1625, 1500, and 1350 RPM. Notice that this motor does not have the wide range between speeds as the consequent pole motor does. Most induction motors would overheat and damage the motor windings if the speed were to be reduced to this extent. This motor, however, has much higher impedance in its windings than most motors. The run windings of most split-phase motors has a wire resistance of 1 to 4 ohms. This motor will generally have a resistance of 10 to 15 ohms in its run winding. It is the high impedance of the windings that permits the motor to be operated in this manner without damage.

Since this motor is designed to slow down when load is added, it is not used to operate high-torque loads. This type of motor is generally used to operate only low-torque loads, such as fans and blowers. The schematic in figure 11-4 shows a multi-speed fan motor and switch.

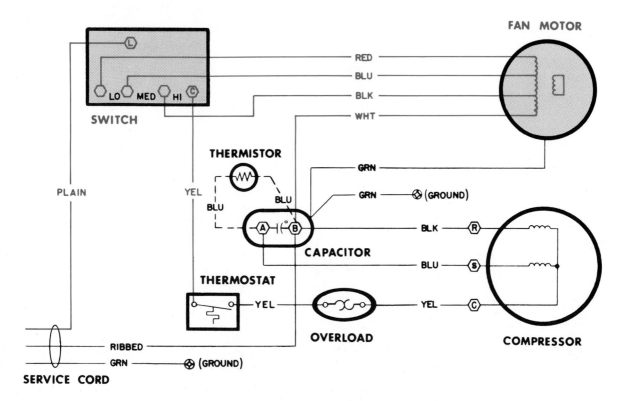

SWITCH POSITION	CONTACTS MADE
LO	L to C, L to LO
MED	L to C, L to MED
HI	L to C, L to HI
OFF	None

FIGURE 11-4 A multi-speed fan motor and switch

REVIEW QUESTIONS

1. Name two ways of changing the speed of a rotating magnetic field.

2. How does the consequent pole motor change speed?

3. Name a disadvantage of the consequent pole motor.

4. Name an advantage of a consequent pole motor.

5. How many steps of speed are common to a multi-speed fan motor?

6. Refer to figure 11-3. Explain what would happen to motor operation if the winding between low and medium should become open.

7. What is an advantage of the multi-speed fan motor over the consequent pole motor?

8. What is a disadvantage of the multi-speed fan motor when compared to the consequent pole motor?

9. How much wire resistance is common for the run winding of most split-phase motors?

10. How much wire resistance is common for the multi-speed fan motor?

SECTION 3

Three-Phase Motors

UNIT 12

Three-Phase Motor Principles

There are three basic types of three-phase motors. These are:

1. The squirrel cage induction motor,
2. The wound rotor induction motor,
3. The synchronous motor.

The type of three-phase motor is determined by the rotor or rotating member. The stator windings for any of these motors is the same. In this unit, the basic principles of operation for three-phase motors will be discussed.

The principle of operation for all three-phase motors is the rotating magnetic field. There are three factors that cause the magnetic field to rotate. These are:

1. The voltages of a three-phase system are 120° out of phase with each other.
2. The three voltages change polarity at regular intervals.
3. The arrangement of the stator windings around the inside of the motor.

Figure 12-1A shows three AC voltages 120° out of phase with each other, and the stator winding of a three-phase motor. The stator illustrates a two-pole, three-phase motor. Two-pole means that there are two poles per phase. AC motors do not generally have actual pole pieces as shown in figure 12-1A, but they will be used here to aid in understanding how the rotating magnetic field is created in a three-phase motor. Notice that pole pieces 1A and 1B

are located opposite each other. The same is true for poles 2A and 2B, and 3A and 3B. The pole pieces 1A and 1B are wound with wire that is connected to phase one of the three-phase system. Notice also that the pole pieces are wound in such a manner that they will always have opposite magnetic polarities. If pole piece 1A has a north magnetic polarity, pole piece 1B will have a south magnetic polarity at the same time.

The windings of pole pieces 2A and 2B are connected to line 2 of the three-phase system. The windings of pole pieces 3A and 3B are connected to line 3 of the three-phase system. These pole pieces are also wound in such a manner as to have the opposite polarity of magnetism.

To understand how the magnetic field rotates around the inside of the motor, refer to figure 12-1B. Notice a line, labeled "A," has been drawn through the three voltages of the system. This line is used to illustrate the condition of the three voltages at this point in time. The arrow drawn inside the motor indicates the greatest strength of the magnetic field at the same point in time. It is to be assumed that the arrow is pointing in the direction of the north magnetic field. Notice in figure 12-1B, that phase 1 is at its maximum positive peak, and that phases 2 and 3 are less than maximum. The magnetic field is, therefore, strongest between pole pieces 1A and 1B.

In figure 12-1C, line B indicates tht the voltage of line 3 is zero. The voltage of line 1 is less

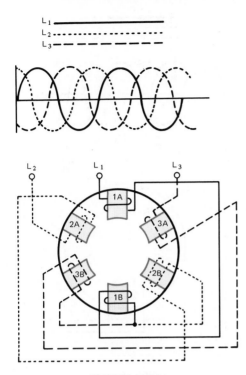

FIGURE 12-1A

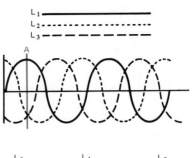
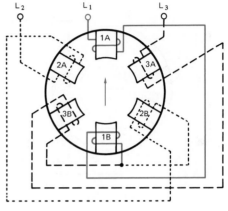

FIGURE 12-1B

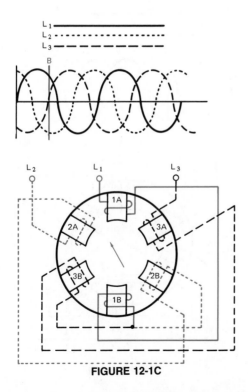

FIGURE 12-1C

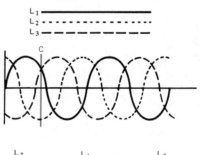
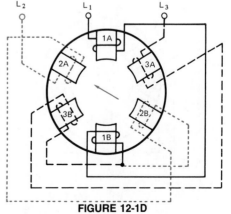

FIGURE 12-1D

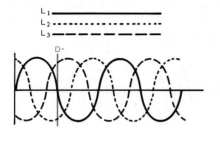

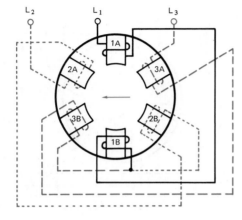

FIGURE 12-1E

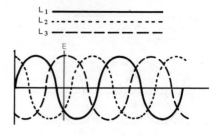

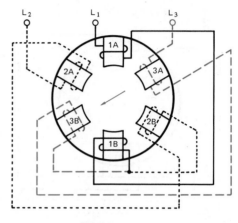

FIGURE 12-1F

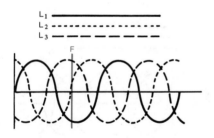

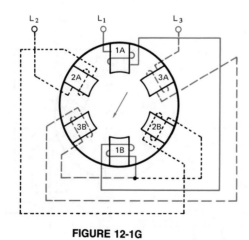

FIGURE 12-1G

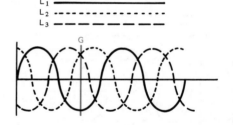

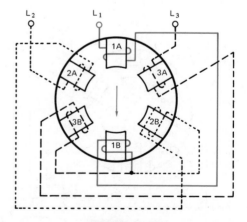

FIGURE 12-1H

SECTION 3   THREE-PHASE MOTORS

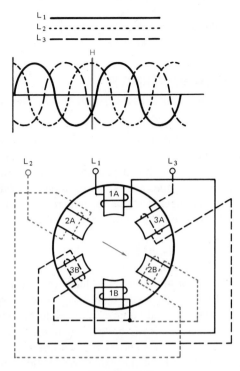

FIGURE 12-1I

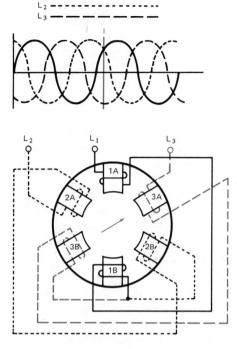

FIGURE 12-1J

than maximum positive; and line 2 is less than maximum negative. The magnetic field at this point is concentrated between the pole pieces of phase 1 and phase 2.

In figure 12-1D, line C indicates that line 2 is at its maximum negative peak and that lines 1 and 3 are less than maximum positive. The magnetic field at this point is concentrated between pole pieces 2A and 2B.

In figure 12-1E, line D indicates that line 1 is zero. Lines 2 and 3 are less than maximum and in opposite directions. At this point, the magnetic field is concentrated between the pole pieces of phase 2 and phase 3.

In figure 12-1F, line E indicates that phase 3 is at its maximum positive peak and lines 1 and 2 are less than maximum and in the opposite direction. The magnetic field at this point is concentrated between pole pieces 3A and 3B.

In figure 12-1G, line F indicates that phase 2 is 0. Line 3 is less than maximum positive; and line 1 is less than maximum negative. The magnetic field at this time is concentrated between the pole pieces of phase 1 and phase 3.

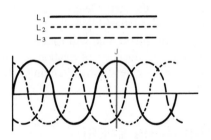

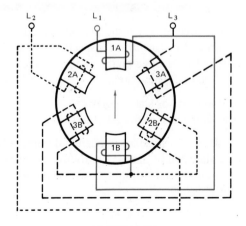

FIGURE 12-1K

In figure 12-1H, line G indicates that phase 1 is at its maximum negative peak; and phase 2 and 3 are less than maximum and in the opposite direction. Notice that the magnetic field is again concentrated between pole pieces 1A and 1B. This time, however, the magnetic polarity is reversed because the current has reversed in the stator winding.

In figure 12-1I, line H indicates phase 2 is at its maximum positive peak and phases 1 and 3 are less than maximum and in the negative direction. The magnetic field is concentrated between pole pieces 2A and 2B.

In figure 12-1J, line I indicates that phase 3 is maximum negative; and phase 1 and 2 are less than maximum in the positive direction. The magnetic field at this point is concentrated between pole pieces 3A and 3B.

In figure 12-1K, line J indicates that phase 1 is at its positive peak; and phases 2 and 3 are less than maximum and in the opposite direction. The magnetic field is again concentrated between pole pieces 1A and 1B. Notice that in one complete cycle of three-phase voltage, the magnetic field has rotated 360° around the inside of the stator winding.

If any two of the stator leads is connected to a different line, the relationship of the voltages will change and the magnetic field will rotate in the opposite direction. The direction of rotation of a three-phase motor can be reversed by changing any two stator leads.

SYNCHRONOUS SPEED

The speed at which the magnetic field rotates is known as the *synchronous* speed. The synchronous speed of a three-phase motor is determined by two factors. These are:

1. The number of stator poles,
2. The frequency of the AC line.

Since 60 Hz is a standard frequency throughout the United States and Canada, the following gives the synchronous speeds for motors with different numbers of poles.

2 Poles	3600 RPM
4 Poles	1800 RPM
6 Poles	1200 RPM
8 Poles	900 RPM

STATOR WINDINGS

The stator windings of three-phase motors are connected in either a wye or delta. Some stators are designed in such a manner as to be connected in either wye or delta, depending on the operation of the motor. Some motors, for example, are started as a wye-connected stator to help reduce starting current, and then changed to a delta connection for running.

Many three-phase motors have dual-voltage stators. These stators are designed to be connected to 240 volts or 480 volts. The leads of a dual-voltage stator use a standard numbering system. Figure 12-2 shows a dual-voltage wye-connected stator. Notice the stator leads have been numbered in a spiral. This diagram shows that numbers 1 and 4 are opposite ends of the same coil. Lead number 7 begins another coil, and this coil is to be connected to the same phase as 1 and 4. Leads 2 and 5 are opposite ends of the same coil. Coil number 8 must be connected with the same phase as leads 2 and 5. Leads 3 and 6 are opposite ends of a coil and must be connected with lead number 9. Keep in mind that figure 12-2 is a schematic diagram,

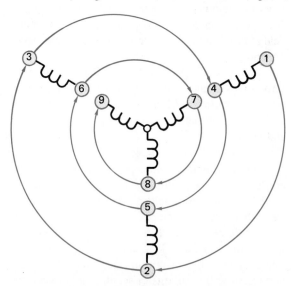

FIGURE 12-2 Numbering a dual-voltage stator

SECTION 3 THREE-PHASE MOTORS

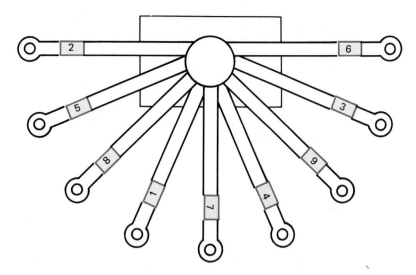

FIGURE 12-3 Leads of a dual-voltage motor

and that when connecting a three-phase motor for operation at the proper voltage, the leads will look more like figure 12-3. This figure illustrates the leads coming out of the terminal connection box on the motor. Some leads are numbered with metal or plastic bands on the wires, and some leads have numbers printed on the insulation of the wire.

Figure 12-4 shows the stator connection for operation on a 480-volt line. Figure 12-5 shows the schematic equivalent of this connection. Notice that the windings have been connected in series. Figure

12-6 shows the stator connection for operation on a 240-volt line. Figure 12-7 shows the schematic equivalent of this connection. When the motor is to be operated on 240 volts, the stator windings are connected in parallel. Notice that leads 4, 5, and 6 are connected together to form another centerpoint. This centerpoint is electrically the same as the point where leads 7, 8, and 9 join together. Figure 12-8 shows the equivalent circuit.

When a motor is operated on a 240-volt line, the current draw of the motor is double the current

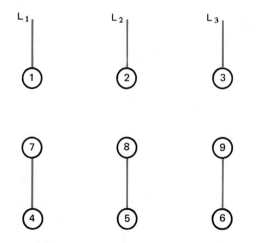

FIGURE 12-4 High-voltage connection

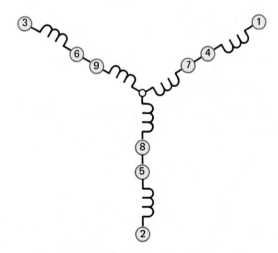

FIGURE 12-5 Windings connected in series

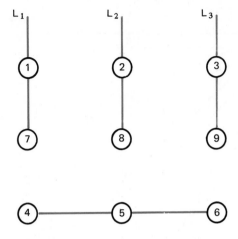

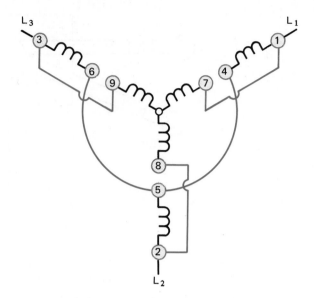

FIGURE 12-6 Low-voltage connection

FIGURE 12-7 Windings connected in parallel

draw of a 480-volt connection. For example, if a motor draws 10 amps of current when connected to 240 volts, it will draw 5 amps when connected to 480 volts. The reason for this is the difference of impedance in the windings between a 240-volt connection and a 480-volt connection. For instance, assume the stator windings of a motor have an impedance of 48 ohms. If the stator windings are connected in parallel as shown in figure 12-9, the total impedance of the windings is 24 ohms.

$$Rt = \frac{R1 \times R2}{R1 + R2}$$

$$Rt = \frac{48 \times 48}{48 + 48}$$

$$Rt = \frac{2304}{96}$$

$$Rt = 24 \text{ ohms}$$

If 240 volts is applied to this connection, 10 amps of current will flow.

$$I = \frac{E}{R}$$

$$I = \frac{240}{24}$$

$$I = 10 \text{ amps}$$

If the windings are connected in series for opera-

tion on a 480-volt line as shown in figure 12-10, the total impedance of the winding is 96 ohms.

$$Rt = R1 + R2$$

$$Rt = 48 + 48$$

$$Rt = 96 \text{ ohms}$$

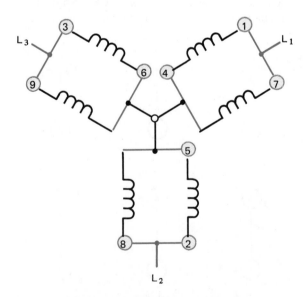

FIGURE 12-8 Equivalent parallel circuit

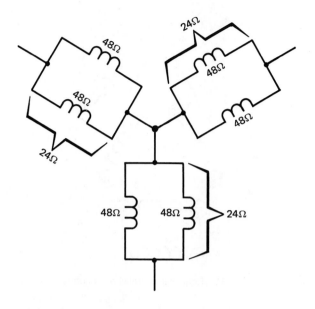

FIGURE 12-9   Total impedance of a parallel connection

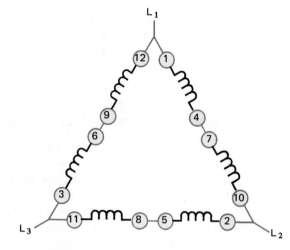

FIGURE 12-11   High-voltage delta connection

DELTA CONNECTIONS

Three-phase motors are also connected in delta. The same standard numbering system is used for delta-connected motors. If a dual-voltage motor is to be connected in delta, there must be 12 leads instead of 9 leads brought out at the terminal box.

Figure 12-11 shows the schematic diagram of a motor connected for operation as a high-voltage delta. Notice that the stator windings for each phase have been connected in series for operation on high

If 480 volts is applied to this winding, 5 amps of current will flow.

$$I = \frac{E}{R}$$
$$I = \frac{480}{96}$$
$$I = 5 \text{ amps}$$

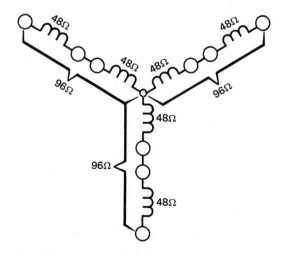

FIGURE 12-10   Impedance adds in series.

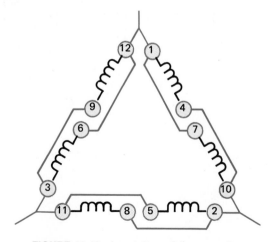

FIGURE 12-12   Low-voltage delta connection

UNIT 12   THREE-PHASE MOTOR PRINCIPLES

voltage. If the motor is to be connected for operation on a low voltage, the windings will be connected in parallel as shown in figure 12-12. Figure 12-13 shows an equivalent parallel connection.

SPECIAL CONNECTIONS

Some three-phase motors designed for operation on voltages higher than 600 volts may have more than 9 or 12 leads brought out at the terminal box. A motor with 15 or 18 leads can be found in high-voltage installations. A 15-lead motor has 3 coils per phase instead of 2. Figure 12-14 shows the proper number sequence for a 15-lead motor. Notice the leads are numbered in the same spiral as a 9-lead motor.

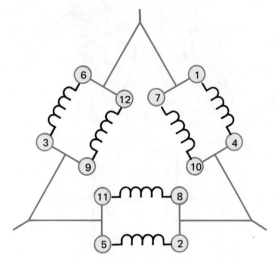

FIGURE 12-13   Equivalent parallel delta connection

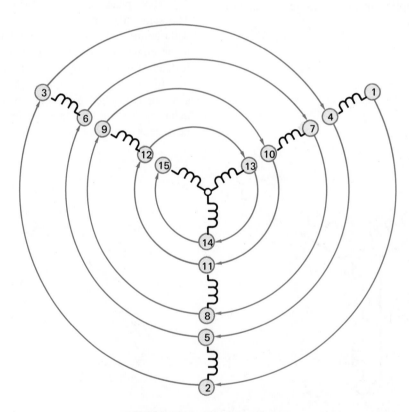

FIGURE 12-14   Fifteen-lead motor

REVIEW QUESTIONS

1. What are the three basic types of three-phase motors?

2. Name three factors that produce a rotating magnetic field.

3. What is synchronous speed?

4. What two factors determine the synchronous speed of a three-phase motor?

5. How is the direction of rotation of a three-phase motor changed?

6. What is the synchronous speed of a four-pole motor when connected to a 60-Hz line?

7. A dual-voltage three-phase motor has a current draw of 50 amps when connected to a 240-volt line. How much current will flow if the motor is connected for operation on 480 volts?

8. If the stator windings of a three-phase motor are connected for operation on high voltage, will the windings be connected in series or parallel?

9. If a dual-voltage motor is connected for operation on low voltage, and the motor is then connected to high voltage, will the motor operate at a faster speed?

10. Why does a dual-voltage motor draw less current when connected to low voltage than it does when connected to high voltage?

UNIT 13

The Squirrel Cage Induction Motor

The squirrel cage induction motor receives its name from the fact that the rotor contains a set of bars that resemble a squirrel cage. If the soft-iron laminations were to be removed from the rotor, it would be seen that the rotor contains a set of metal bars joined together at each end by a metal ring, figure 13-1. Figure 13-2 shows a complete squirrel cage rotor and stator winding.

PRINCIPLE OF OPERATION

The squirrel cage motor is an induction motor. This means that the current flow in the rotor is produced by induced voltage from the rotating magnetic field of the stator. In figure 13-3, a squirrel cage rotor is shown inside the stator winding of a three-phase motor. It will be assumed that the motor shown in figure 13-3 contains four poles per phase, which produces a synchronous speed of 1800 RPM (revolutions per minute) when the stator is connected to a 60-Hz line. When power is first connected to the stator, the rotor is not turning. The magnetic field of the stator cuts the rotor bars at a rate of 1800 RPM. There are three factors that determine the amount of induced voltage. These are:

1. The strength of the magnetic field,
2. The number of turns of wire cut by the mag-

netic field (in the case of a squirrel cage rotor, this will be the number of bars in the rotor),
3. The speed of the cutting action.

Since the rotor is stationary at this time, maximum voltage is induced into the rotor. The induced voltage causes current to flow through the rotor bars. As current flows through the rotor, a magnetic field is produced around each rotor bar. The magnetic field of the rotor is attracted to the magnetic field of the stator, and the rotor begins to turn in the same direction as the rotating magnetic field.

As the speed of the rotor increases, the rotating magnetic field cuts the rotor bars at a slower rate. For example, assume the rotor has accelerated to a speed of 600 RPM. The synchronous speed of the rotating magnetic field is 1800 RPM. Therefore, the rotor is being cut at a rate of 1200 RPM. (1800 RPM − 600 RPM = 1200 RPM). Since the rotor is being cut at a slower rate, less voltage is induced into the rotor. This produces less current flow through the rotor. When the current flow in the rotor is reduced, the current flow in the stator is reduced also.

As the rotor continues to accelerate, the rotating magnetic field cuts the rotor bars at a slower rate. This causes less voltage to be induced into the rotor, and therefore, less current flow in the rotor. Notice that the maximum amount of induced voltage and current occurs when the rotor is not turning at the instant of start. This is the reason

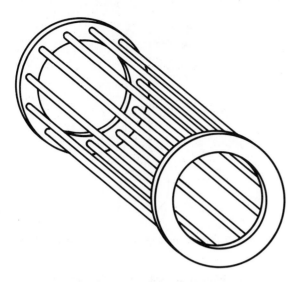

FIGURE 13-1 Squirrel cage bars

FIGURE 13-2 Rotor and stator of a three-phase squirrel cage motor

that AC induction motors require more current to start than to run.

TORQUE

The amount of torque produced by an AC induction motor is determined by three factors. These are:

1. The strength of the magnetic field of the stator,
2. The strength of the magnetic field of the rotor,
3. The phase angle of the current in the stator as compared to the current in the rotor.

Notice that one of the factors that determines the amount of torque produced by an induction motor is the strength of the magnetic field of the rotor. *An induction motor cannot run at synchronous speed.* If the rotor was to accelerate to the speed of the rotating magnetic field, there would be no cutting action of the squirrel cage bars and, therefore, no current flow in the rotor. If there was no current flow in the rotor, there could be no rotor magnetic field and, therefore, no torque.

When an induction motor is operating with no load connected to it, it will run close to the synchronous speed. For example, a four-pole motor that has a synchronous speed of 1800 RPM could run at 1795 RPM at no load. The speed of an AC

induction motor is determined by the amount of torque needed. When the motor is operating at no load, it will produce only the amount of torque needed to overcome its own friction and windage losses. This low torque requirement permits the motor to operate at a speed close to that of the rotating magnetic field.

If a load is connected to the motor, it must furnish more torque to operate the load. This causes the motor to slow down. When the motor speed decreases, the rotating magnetic field cuts the rotor bars at a faster rate. This causes more voltage to be induced in the rotor and, therefore, more current. The increased current flow produces a stronger

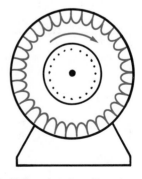

FIGURE 13-3 Voltage is induced into the rotor by the rotating magnetic field.

magnetic field in the rotor, which causes more torque to be produced by the motor. As the current flow increases in the rotor, it causes more current flow to be produced in the stator. This is why motor current will increase as load is added to the motor.

Another factor that determines the amount of torque produced by an induction motor is the phase angle of current in the stator as compared to the current in the rotor. *Maximum torque in an induction motor is produced when the stator current and the rotor current are 90° out of phase with each other*. In an AC circuit, the current flow through a pure resistance is in phase with the voltage. The current flow through a pure inductance, however, is 90° out of phase with the voltage. The current in a pure inductance lags the voltage by 90 electrical degrees.

The stator of an AC motor is inductive. If the rotor could be made resistive, the current of the stator would be 90° out of phase with the current of the rotor, figure 13-4. To understand why maximum torque is produced when these two currents are 90° out of phase with each other, imagine a bolt that must be turned by a wrench as shown in figure 13-5. Maximum turning pressure is applied to the wrench when the force, represented by the arrow, is at a 90° angle to the wrench. If the position of the wrench is changed so that an angle of 120° exists between the wrench handle and the force, less turning pressure is applied to the wrench, figure 13-6. If the position of the wrench is changed as shown in figure 13-7, a 30° angle exists between the force and the wrench handle. The relationship of the current in the stator and the current in the

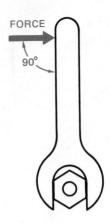

FIGURE 13-5 Maximum force is produced at a 90° angle.

rotor of an induction motor is very similar to the relationship of the applied force and the position of the wrench handle.

CODE LETTERS

Squirrel cage rotors are not all the same. Rotors are made with different types of bars. The type of rotor bars used in the construction of the rotor determines the operating characteristics of the motor. AC motors are given a code letter on their nameplate. The code letter indicates the type of bars used in the rotor. Figure 13-8 shows a rotor with type "A" bars. A type "A" rotor has the highest resistance of any squirrel cage rotor. This means that the starting torque is high since the rotor current is closer to being 90° out of phase with the

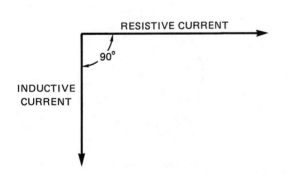

FIGURE 13-4 Resistive current is 90° out of phase with the inductive current.

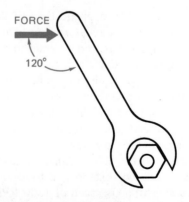

FIGURE 13-6 Force applied at a 120° angle

SECTION 3   THREE-PHASE MOTORS

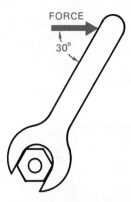

FIGURE 13-7 Force applied at a 30° angle

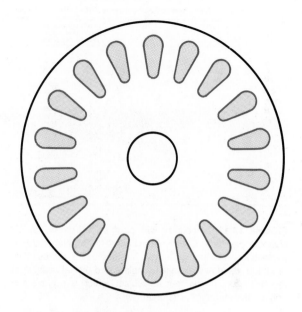

FIGURE 13-9 Type "B-E" rotor

stator current than the other types of rotors. The high resistance of the rotor bars limits the amount of current flow in the rotor when starting. This produces a low starting current for the motor. A rotor with type "A" bars has very poor running characteristics, however. Since the bars are resistive, a large amount of voltage will have to be induced into the rotor to produce an increase in rotor current and, therefore, an increase in the rotor magnetic field. This means that when load is added to the motor, the rotor must slow down a great amount

to produce enough current in the rotor to increase the torque.

Figure 13-9 shows a rotor with bars similar to those found in rotors with code letters B through E. These rotor bars have lower resistance than the

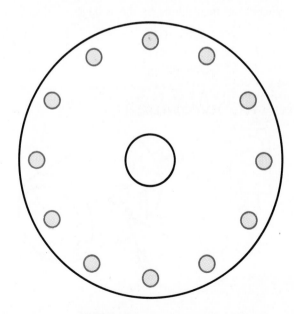

FIGURE 13-8 Type "A" rotor

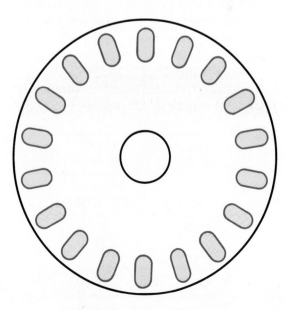

FIGURE 13-10 Type "F-V" rotor

Table 430-7(b). Locked-Rotor Indicating Code Letters

Code Letter		Kilovolt-Amperes per Horsepower with Locked Rotor		
A	....	0	—	3.14
B	....	3.15	—	3.54
C	....	3.55	—	3.99
D	....	4.0	—	4.49
E	....	4.5	—	4.99
F	....	5.0	—	5.59
G	....	5.6	—	6.29
H	....	6.3	—	7.09
J	....	7.1	—	7.99
K	....	8.0	—	8.99
L	....	9.0	—	9.99
M	....	10.0	—	11.19
N	....	11.2	—	12.49
P	....	12.5	—	13.99
R	....	14.0	—	15.99
S	....	16.0	—	17.99
T	....	18.0	—	19.99
U	....	20.0	—	22.39
V	....	22.4	—	and up

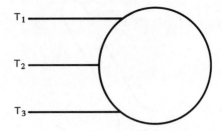

FIGURE 13-13 Schematic symbol of a three-phase squirrel cage induction motor

FIGURE 13-11 (Reprinted with permission from NFPA 70, *National Electrical Code®*, Copyright © 1989, National Fire Protection Association, Quincy, Massachusetts 02269. This reprinted material is not the complete and official position of the NFPA on the referenced subject, which is represented only by the standard in its entirety.)

type "A" rotor. This rotor has fair starting torque, low starting current, and fair speed regulation.

Figure 13-10 shows a rotor with bars similar to those found in rotors with code letters F through V. This rotor has low starting torque, high starting current, and good running torque. This type of rotor also has good speed regulation.

The code letter found on the nameplate is also used to determine the amount of locked rotor current for the motor. Locked rotor current is the amount of current the motor will draw at the moment of starting. Figure 13-11 shows Table 430-7(b) of the National Electrical Code. This table is used to determine the locked rotor current for a squirrel cage rotor. To use the table, the horsepower, code letter, and voltage of the motor must be known. For this example, assume the motor is 10 hp, has a code letter J, and is operated on a 480-volt line. The table lists the locked rotor currents in kilovolt-amperes per horsepower. The table shows that code letter J is 7.1 to 7.99. For this calculation, a midvalue of 7.5 will be used. Since the values are listed in kilovolt-ampere, 7.5 is actually 7500 volt-amperes. To find the locked rotor current, multiply the KVA rating by the horsepower, and then divide by the voltage.

$$I = 7500 \text{ VA} \times 10 \text{ hp}$$
$$I = 75000/480$$
$$I = 156.25 \text{ amps}$$

MOTOR NAMEPLATE	
HP	Phase
10	3
Volts	Amps
240/480	28/14
Hz	FL Speed
60	1745 RPM
Code	SF
J	1.25
Frame	Model No.
XXXX	XXXX

FIGURE 13-12

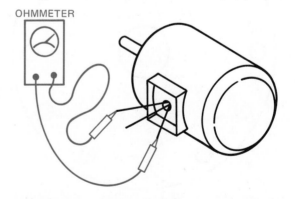

FIGURE 13-14 Testing the stator winding for opens

SECTION 3 THREE-PHASE MOTORS

THE NAMEPLATE

Electric motors have nameplates that give a great deal of information about the motor. Figure 13-12 illustrates the nameplate of a three-phase induction motor. The nameplate shows that the motor is 10 horsepower, it is a three-phase motor and operates on 240 or 480 volts. The full-load running current of the motor is 28 amps when operated on 240 volts, or 14 amps when operated on 480 volts. The motor is designed to be operated on a 60-Hz AC voltage, and has a full-load speed of 1745 RPM. The speed indicates that this motor has four poles per phase. Since the full-load speed is 1745 RPM, the synchronous speed would be 1800 RPM. The motor contains a type J squirrel cage rotor, and has a service factor of 1.25. The service factor is used to determine the amperage rating of the overload protection for the motor. The frame indicates the type of mounting the motor has. Figure 13-13 shows the schematic symbol for a three-phase squirrel cage induction motor.

TESTING THE MOTOR

Most service technicians test a three-phase motor with an ohmmeter. The ohmmeter can be used to check the stator winding for an open condition or a grounded condition. To test the stator winding for an open condition, check the continuity of each winding by measuring the resistance between each of the three windings as shown in figure 13-14. The resistance of each pair of windings should be the same. To test for a grounded

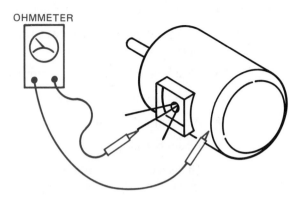

OHMMETER

FIGURE 13-15 Testing the stator winding for grounds

motor, connect one ohmmeter lead to the case of the motor, and the other lead to one of the motor leads. There should be no continuity between any of the leads and the case of the motor, figure 13-15. The ohmmeter, however, will not generally detect a shorted winding. The resistance of the stator windings of most large horsepower motors is so low that they appear under normal conditions to be a short circuit to the ohmmeter. To test for a shorted winding, some method must be used to measure the reactance of the windings instead of their resistance. If the motor will run, an ammeter can be used to measure the current draw of the motor. The current of each line should be equal and within the full-load current rating of the motor. If one line has a higher current reading than the others, it is an indication of a shorted stator winding. If it is not possible to operate the motor, an instrument that measures the actual inductance of the winding can be used. If one winding has a lower inductance than the others, it is shorted.

REVIEW QUESTIONS

1. What three factors determine the amount of torque produced by an AC induction motor?

2. Why does an AC induction motor draw more current when starting than it does when running?

3. Why does the current flow to the motor increase when load is added to the motor?

4. What does the code letter found on the nameplate of the motor indicate?

5. At what degree angle between the stator current and the rotor current is the maximum torque developed?

6. What type of squirrel cage rotor has the highest starting torque?

7. What type of squirrel cage rotor has the best speed regulation?

8. Why can an induction motor never operate at synchronous speed?

9. What does the locked rotor current of a motor indicate?

10. The nameplate of a squirrel cage motor indicates that the motor has a full-load speed of 875 RPM. How many poles per phase does the motor have?

UNIT 14
The Wound Rotor Induction Motor

Another type of three-phase induction motor used for operating large air-conditioning units is the wound rotor induction motor. The stator winding of this motor is the same as the stator of a squirrel cage induction motor. The rotor of the wound rotor motor, however, does not contain squirrel cage bars. The rotor of this motor contains wound coils of wire as illustrated in figure 14-1. The rotor will contain as many poles as there are stator poles. The rotor shown in figure 14-1 would be for a two-pole stator. Notice that there are three separate windings on the rotor. The finish end of each winding is connected together. This forms a wye connection for the rotor winding. The start end of each winding is connected to a separate slip ring located on the rotor shaft.

The slip rings permit the connection of external resistance to the rotor windings. Figure 14-2 shows a schematic diagram of the stator connection and rotor connection of a wound rotor motor. Notice that the wye-connected stator winding is connected directly to the incoming power. The wye-connected rotor is connected to three variable resistors. The dashed line drawn between the resistors indicates that they are mechanically connected together. If the resistance of one is changed, the resistance of the other two changes also.

Resistance is connected to the slip rings by means of carbon brushes as shown in figure 14-3. Since the resistance connection to the rotor is external, the amount of resistance used in the circuit can be controlled. This permits the amount of current flow in the rotor to be controlled. If the current flow in the rotor is limited by the amount of resistance connected in the circuit, the stator current is limited also. A great advantage of the wound rotor motor is that it limits the amount of inrush current when the motor is first started. This eliminates the need for reduced voltage starters or wye-delta starting.

Another advantage of the wound rotor motor is its high starting torque. Since resistors are used to limit current flow in the rotor, the phase angle between the stator current and the rotor current is

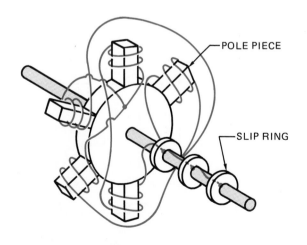

FIGURE 14-1 A wound rotor

close to 90°. The schematic symbol for a wound rotor motor is shown in figure 14-4.

MOTOR OPERATION

When power is applied to the stator winding, a rotating magnetic field is created in the motor. This magnetic field cuts through the windings of the rotor and induces a voltage into them. The amount of voltage induced in the rotor windings is determined by the same three factors that determined the amount of voltage induced in the squirrel cage rotor. The amount of current flow in the rotor is determined by the amount of induced voltage and the amount of resistance connected to the rotor ($I = E/R$). When current flows through the rotor, a magnetic field is produced. This magnetic field is attracted to the rotating magnetic field of the stator.

As the rotor speed increases, the induced voltage decreases because of less cutting action between the rotor windings and rotating magnetic field. This produces less current flow in the rotor and, therefore, less torque. If resistance is reduced, more current can flow, which will increase motor torque, and the rotor will increase in speed. This action continues until the rotor is operating at maximum speed and all resistance has been shorted out of the rotor circuit. When all of the resistance has been shorted out, the motor operates like a squirrel cage induction motor.

STARTING

Most large wound rotor motors use a method of *step starting* as opposed to actual variable resistors. Step starting is similar to shifting the gears in the transmission of an automobile. The transmission is placed in first gear when the car is first started. As the car gains speed, the transmission is shifted to second gear, then third gear, and so on until the car is operating in its highest gear. When a wound rotor motor is step started, it begins with maximum resistance connected in the rotor circuit. As the motor speed increases, resistance is shorted out of the circuit until the windings of the rotor are shorted together. The number of steps can vary from one motor to another, depending on the size of the motor and how smooth a starting action is desired.

There are different control methods used to short out the steps of resistance when starting a wound rotor motor. Some controllers sense the amount of current flow to the stator. This method is known as current limit control. Another method detects the speed of the rotor. This method is known as slip frequency control. One of the most common methods uses time relays to control when resistance is shorted out of the circuit. This method is known as definite time control. Figure 14-5 shows a schematic diagram of a time-controlled starter for a wound rotor motor. In this schematic, the motor circuit is shown at the top of the diagram. A control transformer is used to step the line voltage down

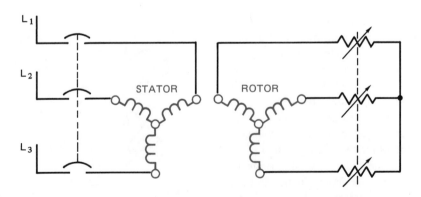

FIGURE 14-2 A wye-connected stator and wye-connected rotor

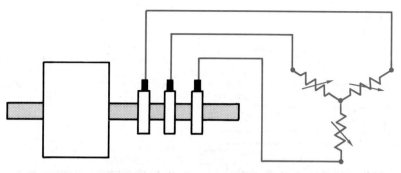

FIGURE 14-3 Resistance is connected to the rotor by brushes and slip rings.

to the value of voltage used in the control circuit. The operation of the circuit is as follows:

1. When the start button is pressed, a circuit is completed through "M" motor starter coil, "TR1" coil, and the overload contact. When "M" coil energizes, all "M" contacts close. The three large load contacts located at the top of the diagram close and connect the stator winding to the line. The "M" contact located beneath the start button is known as the *holding*, *sealing*, or *maintaining* contact. Its job is to provide a continued circuit to the "M" coil when the start button is released. The motor now begins to run in its lowest speed. Maximum resistance is connected in the rotor circuit.

2. "TR1" relay is a timer. For this example, it shall be assumed that all timers are set for a delay of three seconds. When "TR1" coil energizes, it begins a time operation. After three seconds, "TR1" contact closes. This completes a circuit to "S1" coil and "TR2" coil.

3. When "S1" coil energizes, both "S1" contacts close and short out the last three resistors in the

rotor circuit. This causes the motor to accelerate to the next higher speed. When "TR2" coil energizes, it begins timing.

4. At the end of a three-second time period, "TR2" contact closes and completes a circuit to coil "S2" and "TR3."

5. When "S2" coil energizes, both "S2" contacts close and short out the next set of resistors. This permits the motor to accelerate to a higher speed. When "TR3" coil energizes, it begins its timing sequence.

6. After a three second time period, contact "TR3" closes and provides a complete circuit for coil "S3." This causes both "S3" contacts to close and short out the last set of resistors. The motor now accelerates to its highest speed.

7. When the stop button is pressed, the circuit to coil "M" and coil "TR1" is broken. When coil "M" de-energizes, all "M" contacts open. This disconnects the stator winding from the line. When "TR1" coil de-energizes, contact "TR1" opens immediately. This de-energizes coil "S1" and coil "TR2." When coil "S1" de-energizes, both "S1" contacts return to their open position. When coil "TR2" de-energizes, contact "TR2" opens immediately. When contact "TR2" opens, it breaks the circuit to coil "S2" and coil "TR3." When coil "S2" de-energizes, both "S2" contacts reopen. Contact "TR3" opens immediately when coil "TR3" de-energizes. This causes coil "S3" to de-energize and open both "S3" contacts.

8. If the fuse should blow, or the overload contact open, it has the same effect as pressing the stop button.

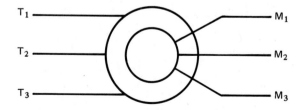

FIGURE 14-4 Schematic symbol of a would rotor induction motor

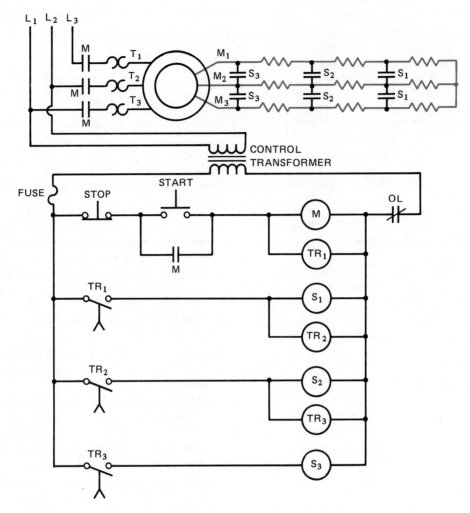

FIGURE 14-5 Definite time starting for a wound rotor motor

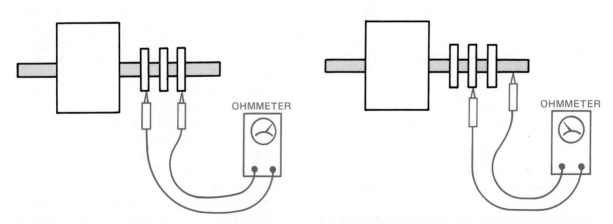

FIGURE 14-6 Testing a rotor for an open winding

FIGURE 14-7 Testing a rotor for a grounded winding

SECTION 3 THREE-PHASE MOTORS

TESTING A WOUND ROTOR MOTOR

Since the stator winding of the wound rotor motor is the same as the squirrel cage motor, the same test procedure can be followed. Testing the rotor of a wound rotor motor is very similar to testing the stator. The rotor can be tested for an open winding with an ohmmeter by checking the continuity between each of the slip rings, figure 14-6. The resistance readings should be the same between each pair of slip rings. To test the rotor for a ground, connect one ohmmeter lead to the shaft, and connect the other lead to each one of the slip rings, figure 14-7. The ohmmeter should show no continuity between the rotor windings and ground. Like the stator winding, the rotor is difficult to test for a shorted winding. To test the rotor for a shorted winding it is generally necessary to use equipment that will measure the inductance of the winding instead of its resistance.

REVIEW QUESTIONS

1. How many slip rings are located on the shaft of the rotor of a wound rotor induction motor?

2. What is the purpose of the slip rings?

3. Name two advantages of the wound rotor motor over the squirrel cage motor.

4. What two factors determine the amount of current flow in the rotor of a wound rotor motor?

5. What does the dashed line drawn between the three resistors shown in figure 14-2 indicate?

6. Why is the starting torque of a wound rotor induction motor higher than the starting torque of a squirrel cage induction motor?

7. The stator of a wound rotor motor has a synchronous speed of 1200 RPM when connected to a 60-Hz line. How many poles per phase are there in the rotor?

8. Refer to figure 14-5. Describe what would happen in this circuit if coil "S1" should be open when the motor started.

9. Refer to figure 14-5. Describe what would happen in this circuit if coil "TR2" should be open when the motor is started.

10. Refer to figure 14-5. Describe what would happen in this circuit if holding contact "M" should become stuck together when the motor is started and not open.

UNIT 15

The Synchronous Motor

The third type of three-phase motor to be discussed is the synchronous motor. This motor has several characteristics that no other type of motor has. Some of the characteristics of a synchronous motor are:

1. The synchronous motor is not an induction motor. This means that it does not depend on induced voltage from the stator to produce a magnetic field in the rotor.
2. The synchronous motor will run at a constant speed from no load to full load.
3. The synchronous motor has the ability to not only correct its own power factor, but can also correct the power factor of other motors connected to the same line.

The synchronous motor has the same type of stator windings as the other two three-phase motors. The rotor of a synchronous motor has windings similar to the wound rotor induction motor, figure 15-1. Notice that the winding in the rotor of a synchronous motor is different, however. The winding of a synchronous motor is one continuous set of coils instead of three different sets as is the case with the wound rotor motor. Notice also that the synchronous motor has only two slip rings on its shaft as opposed to three on the shaft of a wound rotor motor.

STARTING A SYNCHRONOUS MOTOR

The rotor of a synchronous motor also contains a set of type "A" squirrel cage bars. This set of squirrel cage bars is used to start the motor and is known as the *amortisseur* winding, figure 15-2. When power is first connected to the stator, the rotating magnetic field cuts through the type "A" squirrel cage bars. The cutting action of the field induces a current into the squirrel cage bars. The

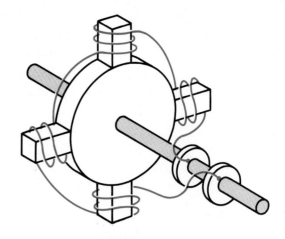

FIGURE 15-1 Rotor

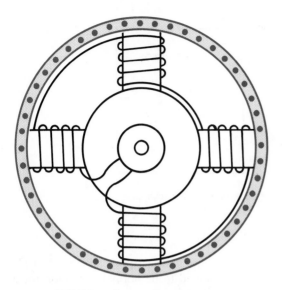

FIGURE 15-2 Amortisseur winding

current flow through the amortisseur winding produces a rotor magnetic field that is attracted to the rotating magnetic field of the stator. This causes the rotor to begin turning in the direction of rotation of the stator field. When the rotor has accelerated to a speed that is close to the synchronous speed of the field, direct current is connected to the rotor through the slip rings on the rotor shaft, figure 15-3. When DC current is applied to the rotor, the windings on the rotor become electromagnets. The electromagnetic field of the rotor locks in step with the rotating magnetic field of the stator. The rotor will now turn at the same speed as the rotating magnetic field. When the rotor begins to turn at the synchronous speed of the field, there is no more cutting action between the field and the amoritisseur winding. This causes the current flow in the amoritisseur winding to cease.

Notice that the synchronous motor starts as a squirrel cage induction motor. Since the rotor bars used are type "A," they have a relatively high resistance, which gives the motor good starting torque and low starting current. A synchronous motor must never be started with DC current connected to the rotor. If DC current is applied to the rotor, the field poles of the rotor become electromagnets. When the stator is energized, the rotating magnetic field begins turning at synchronous speed. The electromagnets of the rotor are attracted to the rotating magnetic field of the stator and are alternately attracted and repelled 60 times a second. As a result, the rotor does not turn.

THE FIELD DISCHARGE RESISTOR

When the stator winding is first energized, the rotating magnetic field cuts through the rotor winding at a fast rate of speed. This causes a large amount of voltage to be induced into the winding of the rotor. To prevent this voltage from becoming excessive, a resistor is connected across the winding. This resistor is known as the field discharge resistor, figure 15-4. It also helps to reduce the voltage induced into the rotor by the collapsing

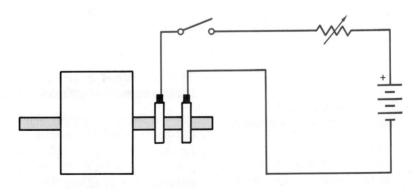

FIGURE 15-3 Direct current is applied to the rotor through the slip rings.

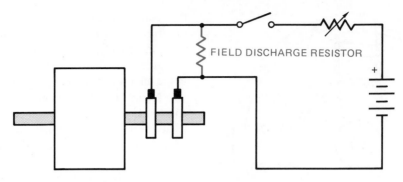

FIGURE 15-4 A field discharge resistor protects the rotor circuit.

magnetic field when the DC current is disconnected from the rotor.

CONSTANT SPEED OPERATION

Although the synchronous motor starts as an induction motor, it does not operate as one. After the amortisseur winding has been used to accelerate the rotor to about 95% of the speed of the rotating magnetic field, direct current is connected to the rotor and the electromagnets lock in step with the rotating field. Notice that the synchronous motor does not depend on induced voltage from the stator field to produce a magnetic field in the rotor. The magnetic field of the rotor is produced by external DC current applied to the rotor. This is the reason that the synchronous motor has the ability to operate at the speed of the rotating magnetic field. As load is added to the motor, the magnetic field of the rotor remains locked with the rotating magnetic field and the rotor continues to turn at the same speed.

POWER FACTOR CORRECTION

The power factor of the synchronous motor can be changed by adjusting the DC excitation current to the rotor. When the DC current is adjusted to the point that the motor current is in phase with the voltage, the motor has a power factor of 100%. This is considered to be normal excitation for the motor. For this example, assume this current to be 10 amps. If the DC power supply is adjusted to a point that the excitation current is less than 10 amps, the rotor is under-excited. This causes the motor to have a lagging power factor like an induction motor. If the excitation current is adjusted above 10 amps, the rotor is over-excited. This causes the motor to have a leading power factor like a capacitor. When a synchronous motor is operated at no load and used for power factor correction, it is generally referred to as a synchronous condenser. Utility companies generally charge industries extra for poor power factor in the plant. For this reason, synchronous motors are often used when a large horsepower motor must be used. Commercial and industrial air-conditioning systems are often the largest single load in a plant or building. It is not uncommon to find synchronous motors being used to operate the compressors of large air-conditioning systems.

THE POWER SUPPLY

The DC power supply of a synchronous motor can be provided by several methods. The most common of these methods is either a small DC generator mounted to the shaft of the motor, or an electronic power supply that converts the AC line voltage to DC voltage.

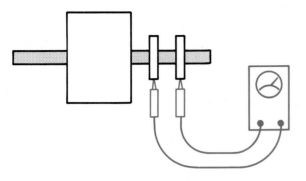

FIGURE 15-5 Testing the rotor for an open winding

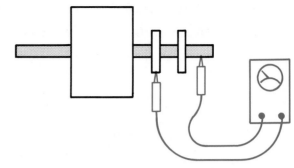

FIGURE 15-6 Testing for rotor for a grounded winding

TESTING THE SYNCHRONOUS MOTOR

The procedure for testing the stator winding of a synchronous motor is the same as that described for testing the stator of a squirrel cage induction motor. The rotor can be tested with an ohmmeter for an open winding or a grounded winding. To test the rotor for an open winding, connect one of the ohmmeter leads to each of the slip rings on the rotor shaft, figure 15-5. Since the rotor winding of a synchronous motor is intended for DC current, the resistance of the wire will be high as compared to the wire resistance of a wound rotor motor. Due to the fact that alternating current flows in the rotor of a wound rotor motor, the current is limited by the inductance of the coil and not its resistance.

To test the rotor for a grounded winding, connect one ohmmeter lead to the shaft of the motor, and the other lead to one of the slip rings. There should be no continuity between the winding and the motor shaft, figure 15-6.

Since the resistance of the rotor is relatively high, a shorted winding can often be found with the ohmmeter. In coils designed for DC current, the resistance of the wire is used to limit the flow of current. For example, assume the specifications of a synchronous motor indicate that the DC excitation voltage should be 125 volts, and that maximum rotor current should be 10 amps. The resistance of the rotor can now be calculated by using Ohm's Law:

$R = E/I$
$R = 125/10$
$R = 12.5$ ohms

If the ohmmeter measures a rotor resistance close to 12.5 ohms, the rotor is good. If the ohmmeter measures a much lower resistance, however, the rotor is shorted.

REVIEW QUESTIONS

1. Name three characteristics of a synchronous motor that the squirrel cage induction motor and the wound rotor motor do not have.

2. What is an amortisseur winding?

3. How many slip rings are located on the shaft of a synchronous motor?

4. How many slip rings are located on the shaft of a wound rotor induction motor?

5. Is a synchronous motor started with DC excitation voltage applied to the rotor?

6. What is the field discharge resistor used for?

7. A synchronous motor has an eight-pole stator. What will be the speed of the rotor when it is under full load?

8. How is it possible to know when a synchronous motor has normal excitation applied to its rotor?

9. How can a synchronous motor be made to have a leading power factor?

10. What is a synchronous condenser?

SECTION 4

Control Components

UNIT 16

Overloads

Overload relays are designed to protect the motor circuit from damage due to overloads. Most overload relays are operated by heat. Since the overload unit must be sensitive to motor current, the heater of the overload relay is connected in series with the motor. In this manner, the amount of current that flows through the motor winding also flows through the overload heater. There are two basic types of overload units used in the air-conditioning field, the solder-melting type and the bi-metal type.

THE SOLDER-MELTING TYPE OF OVERLOAD

The solder-melting type of overload unit is used to a large extent on commercial and industrial air-conditioning units. This type of overload unit contains a rachet wheel that is held stationary by sol-

der. Figure 16-1 illustrates the principle of operation. A serrated wheel is attached to a shaft. The shaft is inserted in a hollow tube. The shaft would be free to rotate inside the tube except for the solder that bonds the two units together.

An electric heating element is wound around the tube as shown in figure 16-2. The heating element is connected in series with the motor, figure 16-3. The current that flows through the motor windings also flows through the heating element. The heating element is calibrated to produce a certain amount of heat when a predetermined amount of current flows through it. As long as the current flowing through it does not exceed a certain amount, there is not enough heat produced to melt the solder. If the motor should become overloaded, an excessive amount of current will flow through the heater and the solder will melt. When the solder melts, the shaft is free to turn.

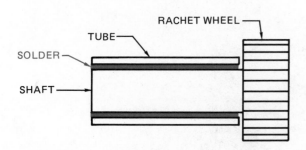

FIGURE 16-1 Shaft is held stationary by solder.

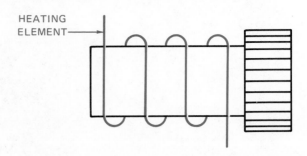

FIGURE 16-2 An electric heating element is wound around the tube.

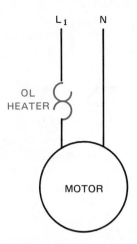

FIGURE 16-3 The overload heater is connected in series with the motor.

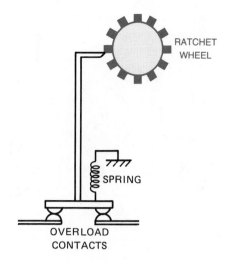

FIGURE 16-4 The ratchet wheel holds the contacts closed.

The rachet wheel is used to mechanically hold a set of spring-loaded contacts closed as shown in figure 16-4. When the solder melts, the rachet wheel is free to turn and the spring causes the contacts to open. The normally closed contacts are connected in series with the coil of the motor starter used to control the motor the overload relay is protecting, figure 16-5. When the overload contacts open, the motor starter coil de-energizes and disconnects the motor from the line.

Notice that this overload relay has two separate sections, the heater section that is connected in series with the motor, and the contact section, which is connected in series with the coil of the motor starter. Notice also that the overload contacts are not used to disconnect the motor from the line. They are used to disconnect the motor starter coil from the line. This type of overload relay has

a set of small auxiliary contacts that are intended to interrupt the current flow in the control circuit only. After the overload has tripped, it must be allowed to cool down enough for the solder to reharden before it can be reset. This is generally true of any type of thermal overload. Figure 16-6 shows a photograph of this type of overload.

BIMETAL TYPE OF OVERLOAD

The bimetal type of overload operates very similarly to the solder-melting type except a bimetal strip is used to cause the contacts to open, figure 16-7. In this unit, the bimetal strip is used to mechanically hold the spring-loaded contacts closed. If the current flow through the heater be-

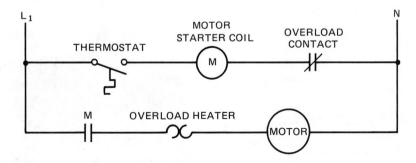

FIGURE 16-5 When the overload contact opens, the motor is disconnected from the line.

FIGURE 16-6 Solder-melting type of overload relay

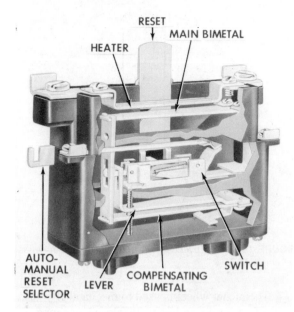

FIGURE 16-8 Bimetal-type overload relay (Courtesy of Furnas Electric Co.)

comes excessive, the bimetal strip will warp and permit the spring to open the contacts. After the overload unit has tripped, the bimetal strip must be allowed some time to cool before it can be reset. This type of unit has an advantage over the solder-melting type in that it can be adjusted for manual reset or automatic reset, figure 16-8. The solder-melting type of overload unit must be manually reset.

Single-phase motors are generally protected with a small automatic reset overload, figure 16-9. These units are constructed in one of two ways. One unit has a small heater connected in series with the motor current, figure 16-10. In this unit the bimetal strip is constructed of a spring metal that provides a snap action when it warps. Notice that the contacts are connected directly to the bimetal strip. This means that the motor current not only flows through the heating element, but also through the bimetal strip. If the motor current becomes excessive, the heater causes the bimetal strip to snap the contacts open and disconnect the motor from the line. Notice that the contacts of this unit

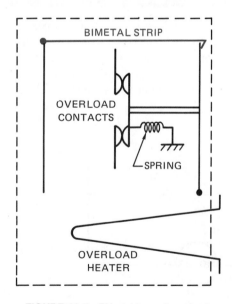

FIGURE 16-7 Bimetal type of overload

FIGURE 16-9 Bimetal overload often used on fractional horsepower single-phase motors

106

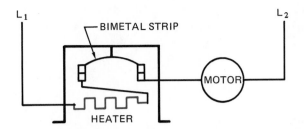

FIGURE 16-10   Small overload unit with heater

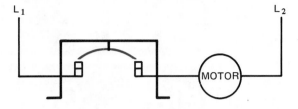

FIGURE 16-11   Small overload unit without heater

are used to interrupt the motor current. When the bimetal strip has cooled enough, it snaps back to its original position and recloses the contacts.

The second type of small overload unit does not contain a heating element, figure 16-11. In this type of unit, the bimetal strip is used as the heating element. As current flows through the bimetal strip, it begins to heat. If motor current does not become excessive, the bimetal strip does not become heated enough to cause the contacts to open. If the current does become excessive, however, the contacts snap open and disconnect the motor from the line.

These overload units can be tested with an ohmmeter for a complete circuit. If the ohmmeter indicates no continuity through them when they are cool, they are defective and must be replaced. Care must be taken to replace these units with the correct size. Overload units are designed to open their contacts when the motor current reaches 115% to 125% of full-load current. The exact rating is determined by the national electrical code. If an overload unit of too small a rating is installed, it will trip when there is no overload on the motor. If an overload unit of too high a value is used, the motor may be destroyed before the overload contacts open and disconnect the motor.

Notice that all of the overload units discussed are operated by sensing heat. For this reason, a heavy motor overload will cause more heat production and the unit will trip faster than it will under a light overload. Another factor that can affect these units is ambient air temperature. Overload relays will trip faster in hot weather than they will in cool weather. In certain parts of the country, it is often necessary to replace the heater elements of industrial-type overload units to match the season. In winter it may be desirable to use a slightly smaller heater element than normal, and in summer it may be necessary to use a slightly larger heater.

REVIEW QUESTIONS

1. What are the two basic types of industrial overload units?

2. What is the advantage of the bimetal type of industrial overload unit?

3. Industrial overload units are divided into two sections. What are they?

4. At what percentage of full-load motor current are overload units generally set to trip?

5. When using an industrial type of overload unit, what are the contacts connected in series with?

6. What is the difference between the two types of small overload units?

7. In the small overload unit which does not contain a heater, what is used to sense the current flow through the motor?

UNIT 17

Relays, Contactors, and Motor Starters

The relay is a magnetically-operated switch. This switch, however, can have multiple sets of contacts, and the contacts can be open or closed. The advantage of the relay is control. A single pilot device can be used to control the input or coil of the relay, and the output or contacts can control several different devices. An example of this is shown in the circuit of figure 17-1. A flow switch is used to control the coil of a magnetic relay. When the flow switch closes, the coil of relay FSCR (Flow Switch Control Relay) is connected to the line. When current flows through the coil, the relay is energized, and all FSCR contacts change position. No-

tice that one FSCR contact is connected in series with the compressor motor. This contact does not actually start the motor, but it permits the thermostat to control the motor. This particular type of control is known as *interlocking*. Interlocking is used to prevent some function from happening until some other function has occurred. In this case, the thermostat can not start the compressor until there is airflow in the system.

The second FSCR contact is normally open. When FSCR coil energizes, this contact closes and turns on a green pilot light to indicate there is airflow in the system. The third FSCR contact is nor-

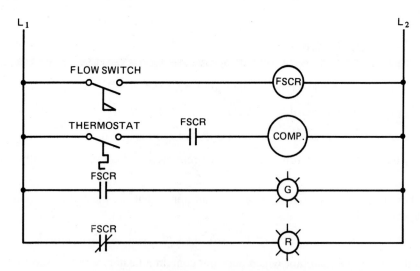

FIGURE 17-1 One relay controls several devices

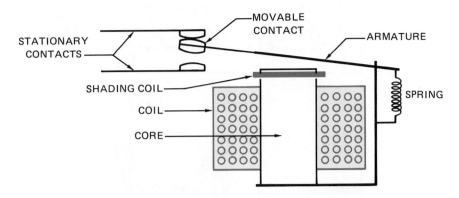

FIGURE 17-2 Simple relay

mally closed. It is used to turn off a red pilot light, which indicates there is no airflow in the system.

PRINCIPLE OF OPERATION

The relay operates on the *solenoid* principle. A solenoid is an electrical device that converts electrical energy to linear motion. This principle is illustrated in figure 17-2. A coil of wire is wound around an iron core. When current flows through the coil, a magnetic field is developed in the iron core. The magnetic field of the iron core attracts the movable arm, known as the armature, and overcomes the strength of the spring holding the arm away from the iron core. Notice that a movable contact is connected to the armature. In its present position, the movable contact makes connection with a stationary contact. This contact set is normally closed. When the armature is attracted to the iron core, the movable contact breaks connection with one stationary contact and makes connection with another. This relay has both a normally open and normally closed set of contacts. Notice that the movable contact is common to both of the stationary contacts. The movable contact would be the common and the stationary contacts would be labeled normally open and normally closed. A schematic of this type of relay is shown in figure 17-3. This illustration shows a relay with only one set of contacts. In practice, it is common to find this type of relay with several sets of contacts.

Notice in figure 17-2 that a shading coild has been added to the iron core. The shading coil is used with AC relays to prevent contact chatter and hum. The shading coil operates in the same way it does in the shaded-pole motor. It opposes a change of magnetic flux. The shading coil is used to provide a continuous magnetic flow to the armature when the voltage of the AC waveform is zero. DC-operated relays do not contain a shading coil since the magnetic flux is constant.

Another type of relay is shown in figure 17-4. This relay uses a plunger-type of solenoid. Notice that the coil is surrounded by the iron core. There is an opening in the iron core through which the shaft of the armature can pass. When the coil is energized, the armature is attracted to both ends

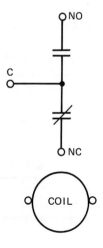

FIGURE 17-3 Schematic of a simple relay

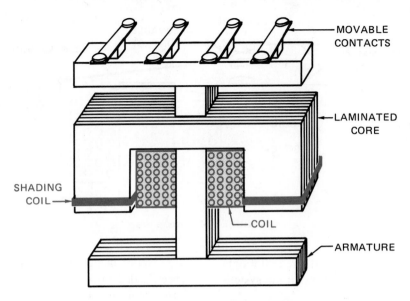

FIGURE 17-4 Plunger type of solenoid

of the core. This creates a stronger magnetic field than the relay discussed in figure 17-2. Notice the shading coils around both ends of the core. Notice also that the core and armature are constructed of laminated sheets. The core and armature are laminated to help prevent the induction of *eddy currents* into the core. Eddy currents are currents induced in the core material by the magnetic field of the coil. Eddy currents are generally unwanted because they heat the core and cause a power loss.

The plunger-type of solenoid is generally used with relays that use double-break contacts. A double-break contact is one that breaks connection at two points as shown in figure 17-5. Notice there are two stationary contacts and one movable contact. The movable contact is used to bridge the gap between the two stationary contacts. This type of contact arrangement is preferred for relays that must control high voltage and current. Notice that the

surface of the contact is curved. This curved surface provides a wiping action when the contacts make connection. The wiping action helps to keep contact surfaces clean. Contact surfaces should never be filed flat. This would permit oil and dirt to collect on the surface of the contact and cause poor connection.

CONTACTORS AND MOTOR STARTERS

The term RELAY is often used to describe any type of magnetically-operated switch. A relay is actually a control device that contains small auxiliary contacts designed to operate only low-current loads.

A contactor is very similar to a relay except that a contactor contains large-load contacts designed to control large amounts of current. In the heating and air-conditioning field, contactors are often used to connect power to resistance heater banks. A photograph of a contactor is shown in figure 17-6. Contactors may contain auxiliary contacts as well as load contacts.

Motor starters are basically contactors with the

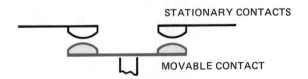

FIGURE 17-5 A set of double-break contacts

STATIONARY CONTACTS

MOVABLE CONTACT

FIGURE 17-6 Contactor

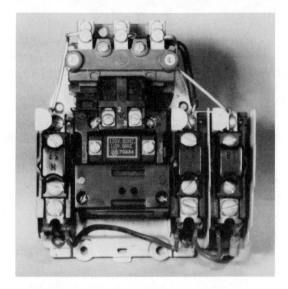

FIGURE 17-7 Motor starter with overload relays

addition of overload relays. Motor starters generally contain auxiliary contacts as well as load contacts. The auxiliary contacts are used as part of the control circuit, and the load contacts are used to connect the motor to the line. A photograph of a motor starter is shown in figure 17-7.

REVIEW QUESTIONS

1. What is a solenoid?

2. What type of relays contain a shading coil?

3. What purpose does the shading coil serve?

4. What is the movable part of a relay called?

5. Why is the core material of a relay laminated?

6. What are eddy currents?

7. What effect do eddy currents have on a relay?

8. Why are contact surfaces curved?

9. What is the difference between a relay and a contactor?

10. What is the difference between a contactor and a motor starter?

UNIT 18
The Solid-State Relay

The solid-state relay is a device that has become increasingly popular for switching applications. The solid-state relay has no moving parts, it is resistant to shock and vibration, and is sealed against dirt and moisture. The greatest advantage of the solid-state relay, however, is the fact that the control input voltage is isolated from the line device the relay is intended to control. Refer to figure 18-1.

Solid-state relays can be used to control either a DC load or an AC load. If the relay is designed to control a DC load, a power transistor is used to connect the load to the line as shown in figure 18-2. The relay shown in figure 18-2 has a *light-emitting diode* (LED) connected to the input or control voltage. When the input voltage turns the LED on, a photo detector connected to the base of the transistor turns the transistor on and connects the load to the line. This optical coupling is a very common method used with solid-state relays. The relays that use this method of coupling are referred to as being *opto-isolated,* which means the load

side of the relay is optically isolated from the control side of the relay. Since a light beam is used as the control medium, no voltage spikes or electrical noise produced on the load side of the relay can be transmitted to the control side of the relay.

Solid-state relays intended for use as AC controllers have a triac connected to the load circuit in place of a power transistor. Refer to figure 18-3. In this example, an LED is used as the control device just as it was in the previous example. When the photo detector "sees" the LED, it triggers the gate of the triac and connects the load to the line.

Although opto-isolation is probably the most common method used for the control of a solid-state relay, it is not the only method used. Some relays use a small reed relay to control the output. Refer to figure 18-4. A small set of reed contacts are connected to the gate of the triac. The control circuit is connected to the coil of the reed relay. When the control voltage causes a current to flow through the coil, a magnetic field is produced around

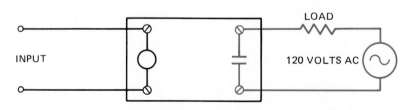

FIGURE 18-1 Solid-state relay

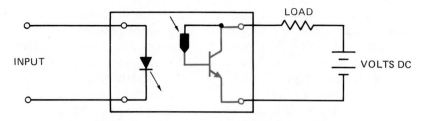

FIGURE 18-2 Power transistor used to control DC load

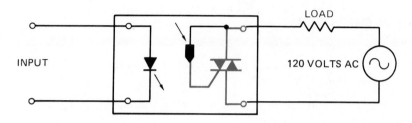

FIGURE 18-3 Triac used to control an AC load

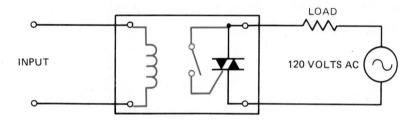

FIGURE 18-4 Read relay controls the output.

the coil of the relay. This magnetic field closes the reed contacts, which causes the triac to turn on. In this type of solid-state relay, a magnetic field is used to isolate the control circuit from the load circuit instead of a light beam.

The control voltage for most solid-state relays ranges from about 3 to 32 volts and can be DC or AC. If a triac is used as the control device, load voltage ratings of 120 to 240 VAC are common and current ratings can range from 5 to 25 amps. Many solid-state relays have a feature known as *zero switching*. Zero switching means that if the relay is told to turn off when the AC voltage is in the middle of a cycle, it will continue to conduct until the AC voltage drops to a zero level and then turn off. For example, assume the AC voltage is at its positive peak value when the gate tells the

FIGURE 18-5 Solid-state relays (Courtesy of International Rectifier)

UNIT 18   THE SOLID-STATE RELAY

triac to turn off. The triac will continue to conduct until the AC voltage drops to a zero level before it actually turns off. Zero switching can be a great advantage when used with some inductive loads such as transformers. The core material of a transformer can be left saturated on one end of the flux swing if power is removed from the primary winding when the AC voltage is at its positive or negative peak. This can cause inrush currents of up to 600% of the normal operating current when power is restored to the primary.

Solid-state relays are available in different case styles and power ratings. Figure 18-5 shows a typical solid-state relay. Some solid-state relays are designed to be used as time-delay relays. One of the most common uses for the solid-state relay is the I/O (eye-oh) track of a programmable controller, which is to be covered in a later unit.

REVIEW QUESTIONS

1. What electronic component is used to control the output of a solid-state relay used to control a DC voltage?

2. What electronic component is used to control the output of a solid-state relay used to control an AC voltage?

3. Explain opto-isolation.

4. Explain magnetic isolation.

5. What is meant by zero switching?

UNIT 19

The Control Transformer

The transformer is a device that has the ability to change the value of AC voltage and current without a change of frequency. Most of the transformers used in the air-conditioning and refrigeration field are known as isolation transformers. This means that the primary and secondary windings are magnetically coupled but electrically isolated from each other. Figure 19-1 illustrates the basic principle of operation of a transformer. This transformer contains two separate windings, the primary and the secondary. The primary is the winding which is connected to the power source and brings power to the transformer. The secondary winding is used to supply power to the load. Notice that there is no electrical connection between the two windings. If one lead of an ohmmeter is connected to one of the primary leads and the other ohmmeter lead is connected to one of the secondary leads,

the ohmmeter should indicate no continuity between the two windings.

PRINCIPLE OF OPERATION

The transformer operates by magnetic induction. When current flows through the primary winding, a magnetic field is created in the winding. Since the secondary winding is wound on the same core as the primary, the magnetic field of the primary induces a voltage into the secondary. This action is known as *mutual induction*. The amount of voltage induced into the secondary is determined by the ratio of the number of turns of wire in the primary as compared to the number of turns of wire in the secondary. For example, assume that the primary winding shown in figure 19-1 contains

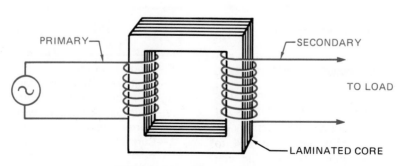

FIGURE 19-1   A basic transformer

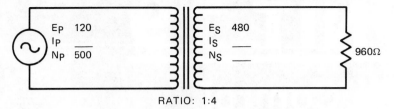

FIGURE 19-2 Schematic for transformer

120 turns of wire and is connected to 120 volts AC. This means that each turn of the primary has a voltage drop of 1 volt. If the secondary winding also has 120 turns of wire, and 1 volt is induced into each turn, then the output voltage of the secondary is 120 volts also. This transformer has a turns ratio of 1:1, which is to say that the primary contains 1 turn of wire for each turn of wire in the secondary.

Now assume that the number of turns of wire in the secondary has been changed to 60. If the number of turns in the primary has not been changed, there is still 1 volt for each turn of wire. This will produce a secondary voltage of 60 volts ($60 \times 1 = 60$).

If the number of turns of wire on the secondary is changed to 240, the output voltage of the secondary will be 240 volts ($240 \times 1 = 240$). Notice that the transformer has the ability to increase or decrease the amount of the secondary voltage. If the voltage of the secondary is less than the primary voltage, the transformer is known as a step-down transformer. If the secondary voltage is greater than the primary voltage, it is known as a step-up transformer.

VOLTAGE AND CURRENT RELATIONSHIPS

It would first appear that the transformer has the ability to give more than it receives. This is not the case, however. Transformers are extremely efficient devices; they generally operate at 95% to 98% efficiency. For this reason, when working with transformers it is generally assumed that the power out of the transformer is equal to the power being put into the transformer.

Figure 19-2 shows the schematic symbol for a transformer. The primary has been connected to 120 volts. The secondary has a voltage of 480 volts and is connected to a load resistor of 960 ohms. This transformer has a turns ratio of 1:4, which is to say that there is 1 turn of wire in the primary for every 4 turns of wire in the secondary. The amount of current in the secondary (I_s) can be computed by using Ohm's Law.

$$I = E/R$$
$$I = 480/960$$
$$I = .5 \text{ amp}$$

Question: If the secondary of this transformer has a current flow of .5 amp, how much current is required in the primary? There are actually several methods that can be used to solve this problem. The most accepted method is to use the formulas shown in figure 19-3. Since both the primary and secondary voltage is known, the formula that contains voltage and current will be used to solve the problem.

$$\frac{E_p}{E_s} = \frac{N_p}{N_s} \qquad \frac{E_p}{E_s} = \frac{I_s}{I_p} \qquad \frac{N_p}{N_s} = \frac{I_s}{I_p}$$

E_p—Voltage of the primary
E_s—Voltage of the secondary
N_p—Number of turns of wire in the primary
N_s—Number of turns of wire in the secondary
I_p—Current flow in the primary
I_s—Current flow in the secondary

FIGURE 19-3 Transformer formulas

$$\frac{E_p}{E_s} = \frac{I_s}{I_p}$$

$$\frac{120}{480} = \frac{.5}{I_p}$$

$$120\, I_p = 240$$

$$I_p = 2$$

Notice that the transformer must have a current draw of 2 amps on the primary to supply a current of .5 amps at the secondary. If the power (volts × amps) is computed for both the primary and the secondary, it will be seen that they are equal.

Primary	Secondary
120 × 2 = 240	480 × .5 = 240

The number of turns of wire can now be computed.

$$\frac{E_p}{E_s} = \frac{N_p}{N_s}$$

$$\frac{120}{480} = \frac{500}{N_s}$$

$$120\, N_s = 240{,}000$$

$$N_s = 2000$$

Notice that the secondary has 2000 turns of wire compared to 500 turns in the primary. This is consistent with the turns ratio, which states there is 1 turn of wire in the primary for every 4 turns in the secondary.

The transformer shown in figure 19-4 is a step-down transformer, which has a primary voltage of 120 volts and a secondary voltage of 24 volts. The secondary is connected to a load resistance of 6 ohms. The current flow in the secondary winding is:

$$I = E/R$$
$$I = 24/6$$
$$I = 4 \text{ amps}$$

Now that the secondary current is known, the amount of primary current can be computed.

$$\frac{E_p}{E_s} = \frac{I_s}{I_p}$$

$$\frac{120}{24} = \frac{4}{I_p}$$

$$120\, I_p = 96$$

$$I_p = .8 \text{ amp}$$

If the amount of power for both the primary and the secondary is computed, it will be seen that they are the same.

Primary	Secondary
120 × .8 = 96	24 × 4 = 96

The number of turns of wire in the secondary can now be computed.

$$\frac{E_p}{E_s} = \frac{N_p}{N_s}$$

$$\frac{120}{24} = \frac{500}{N_s}$$

$$120\, N_s = 12000$$

$$N_s = 100$$

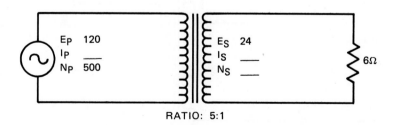

RATIO: 5:1

FIGURE 19-4 Step-down transformer

UNIT 19 THE CONTROL TRANSFORMER

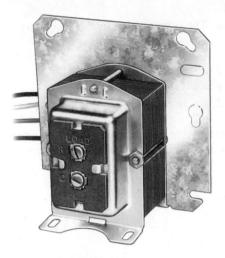

FIGURE 19-5 A 24-volt transformer (Courtesy of Honeywell Inc.)

Notice that the number of turns of wire in the secondary, 100, as compared to the turns of wire in the primary, 500, is consistent with the turns ratio of 5:1.

RESIDENTIAL CONTROL TRANSFORMERS

Control transformers are used to change the value of line voltage to the value needed for the control circuit. Most residential air-conditioning systems operate on a control voltage of 24 volts AC. The amount of current needed will vary from one system to another, but it is generally less than 1 amp. The primary voltage for residential control transformers is 120 or 240 volts. A photograph of a control transformer used in residential applications is shown in figure 19-5. The primary lead wires for most of these transformers will be black in color. The color of the secondary leads will vary from one manufacturer to another.

INDUSTRIAL CONTROL TRANSFORMERS

Most industrial and commercial air-conditioning systems operate on 240 or 480 volts. The con-

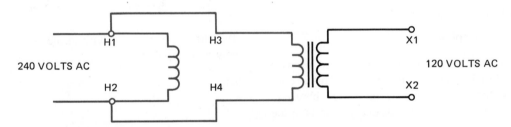

FIGURE 19-6 Primaries connected in parallel for 240-volt operation

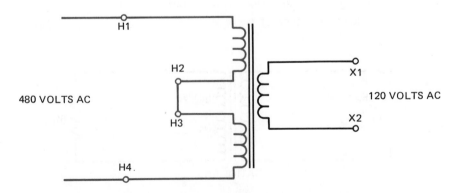

FIGURE 19-7 Primaries connected in series for 480-volt operation

SECTION 4 CONTROL COMPONENTS

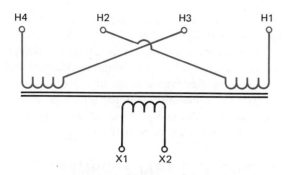

FIGURE 19-8 Primary leads are crossed.

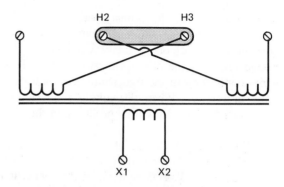

FIGURE 19-10 Metal link used to make a 480-volt connection

trol voltage for most of these units is 120 or 24 volts AC. Most industrial control transformers contain two primary windings and one or two secondary windings. In the following explanation, it will be assumed that the transformer has two primary windings and one secondary winding. In this type of transformer, each primary winding has a voltage rating of 240 volts, and the secondary winding has a voltage rating of 120 volts. There is a turns ratio of 2:1 between each of the primary windings and the secondary winding.

There is a standard for marking the terminals of control transformers. One of the primary windings will be identified with terminal markings of H1 and H2. The other primary winding will be identified with terminal markings of H3 and H4. The secondary winding will be identified with terminal markings of X1 and X2.

If the transformer is to be used to change a primary voltage of 240 volts into 120 volts, the two primary windings will be connected in parallel

as shown in figure 19-6. Since the two primary windings are connected in parallel, each will receive the same voltage. This will produce a turns ratio of 2:1 between the primary windings and the secondary windings. If 240 volts is connected to the primary of a 2:1 ratio transformer, the secondary voltage will be 120 volts.

If the transformer is to be used to change 480

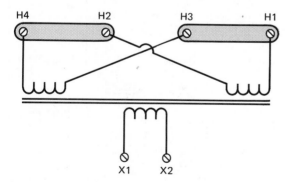

FIGURE 19-9 Metal links used to make a 240-volt connection

FIGURE 19-11 Industrial control transformer

UNIT 19   THE CONTROL TRANSFORMER

volts to 120 volts, the primary windings will be connected in series as shown in figure 19-7. In this connection, H2 of one primary winding is connected to H3 of the other primary winding. This series connection of the two windings produces a turns ratio of 4:1. When 480 volts is connected to the primary, 120 volts will be produced in the secondary.

The primary windings of most control transformers have leads H2 and H3 crossed as shown in figure 19-8. This is done to aid in the connection of the primary. For example, if it is desired to operate the transformer with the primary windings connected in parallel, a metal link is used to connect leads H1 and H3 together. Another metal link is used to connect leads H2 and H4 together, figure

19-9. Compare this lead connection with the schematic shown in figure 19-6.

If it is desired to connect the primary windings for operation on 480 volts, terminals H2 and H3 are joined together with a metal link, figure 19-10. Compare this connection with the schematic shown in figure 19-7. A photograph of an industrial control transformer is shown in figure 19-11.

TESTING THE TRANSFORMER

An ohmmeter is generally used to test the windings of the transformer. To test the transformer, check for continuity through each set of windings. For example, there should be continuity

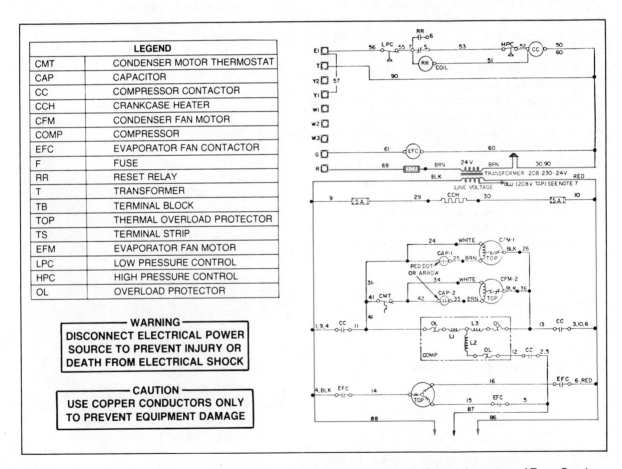

FIGURE 19-12 Control transformer used to provide low voltage for the control circuit (Schematic courtesy of Trane Corp.)

between leads H1 and H2; H3 and H4; and X1 and X2. There should be no continuity between any of the windings such as H1 and H3, or H1 and X1. Also check for a grounded winding by testing to be sure there is no continuity between any of the windings and the case of the transformer.

The output voltage of the transformer should be tested with an AC voltmeter. If the output voltage is not close to the rated voltage, the transformer is probably defective. If the transformer is tested without a load connected to the secondary, it is normal for the secondary voltage to be slightly higher than the rated voltage. For example, a 24-volt transformer may have an output voltage as high as 28 volts without load connected to it. The voltage rating of the transformer assumes it is supplying full-rated current to the load. If the voltage is tested when the transformer is under full load, the rated voltage should be seen. Notice the use of the control transformer in the schematic shown in figure 19-12.

REVIEW QUESTIONS

1. What is an isolation transformer?

2. Define a step-up transformer.

3. Define a step-down transformer.

4. The primary of a transformer is connected to 120 volts AC. The secondary has a voltage of 30 volts and is connected to a resistance of 5 ohms. How much current will flow in the primary of the transformer?

5. What is the amount of control voltage used in most residential air-conditioning systems?

6. What is the amount of control voltage used in most industrial air-conditioning systems?

7. What is the color of the primary leads of most control transformers used for residential service?

8. How many primary windings are generally contained in an industrial control transformer?

9. What is the turns ratio of each of these primary windings as compared to the secondary winding?

10. When an industrial control transformer is to be operated on 480 volts, are the primary windings connected in parallel or series?

UNIT 20

Starting Relays

When a split-phase motor is started, it is often necessary to disconnect the start windings when the motor reaches about 75% of full speed. In an open case motor, this job is generally done by the centrifugal switch. Some single-phase motors are hermetically sealed, however, and a centrifugal switch cannot be used. When this is the case, a starting relay must be used. A starting relay is located away from the motor and is used to disconnect the start windings when the motor has reached about 75% of its full speed. There are three basic types of starting relays in general use:

1. The hot-wire relay
2. The current relay
3. The potential relay

THE HOT-WIRE RELAY

The hot-wire relay is so named because it uses a length of resistive wire connected in series with the motor to sense motor current. A diagram of this type of relay is shown in figure 20-1. When the thermostat contact closes, current can flow from line 1 to terminal L of the relay. Current then flows through the resistive wire, the movable arm, and the normally closed contacts to the run and start windings. When current flows through the resistive wire, its temperature increases. This increase of temperature causes the wire to expand in length.

When the length of the resistive wire increases, the movable arm is forced to move down. As the arm moves down, tension is applied to the springs of both contacts. The relay is so designed that the start contact will snap open first. When the start winding is disconnected, the current flow to the motor will decrease. If the motor current is not excessive, the resistive wire will not expand enough to cause the run contact to open. If the current flow is excessive, however, the wire will continue to expand and the contact connected in series with the run winding will open.

Notice that this type of relay is used as both a starting relay and an overload relay. One disadvantage of the hot-wire relay is that it must be permitted to cool after each operation. A motor using this type of starting relay cannot be started in rapid succession. Figure 20-2 shows a photograph of the hot-wire type of starting relay.

Testing this relay is difficult. An ohmmeter can be used to check for continuity between the L terminal and the start and main winding terminals. To properly test this relay, an ammeter should be used to make certain the start contact opens and disconnects the start winding. If the relay is opening on overload, the ammeter can be used to check the current draw of the motor. This will determine if the motor is actually overloaded, or if the relay is opening when it should not. A good rule to follow concerning starting relays is to always test them if the motor has been damaged. It makes poor busi-

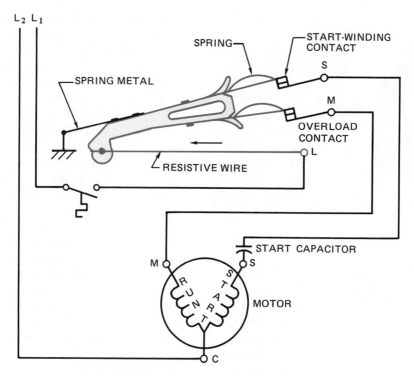

FIGURE 20-1 Hot-wire relay connection

ness sense to damage a new motor because of not checking the starting relay.

When replacing this relay, it is necessary to use the correct replacement. Since the relay is op-

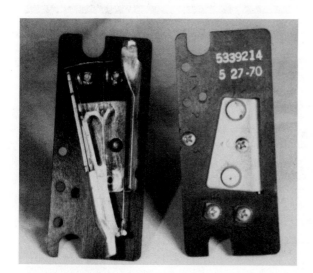

FIGURE 20-2 Hot-wire type of starting relay

erated by motor current, it has been designed to open its contacts when a specific amount of current flows through the circuit. The relay must, therefore, be matched to the characteristics of the motor it is intended to control.

THE CURRENT RELAY

The current relay also operates by sensing the amount of current flow in the circuit. This type of relay operates on the principle of a magnetic field instead of expanding metal. The current relay contains a coil of a few turns of large wire, and a set of normally open contacts, figure 20-3. The coil of the relay is connected in series with the run winding of the motor as shown in figure 20-4. The contacts are connected in series with the start winding. When the thermostat contact closes and connects power to the motor, the starting contacts of the relay are open. Since no power is applied to the start winding, the motor cannot start. This causes a current of about three times the normal full-load cur-

FIGURE 20-3 Current type of starting relay

rent to flow in the run winding. The high current flow through the coil of the start relay produces a strong magnetic field. The magnetic field is strong enough to cause the solenoid to close the starting contacts. When the starting contacts close, power is applied to the start winding and the motor begins to turn. As the motor accelerates, the current flow through the run winding decreases rapidly. When the current flow through the relay coil decreases, the strength of the magnetic field becomes weaker. When the motor has reached about 75% of full speed, the magnetic field is weak enough to permit the solenoid to reopen the starting contacts. This disconnects the start winding from the circuit and the motor continues to operate normally.

Notice that the current relay is used to disconnect the start windings only and does not pro-

vide overload protection. A motor using this type of starting relay must be provided with separate overload protection.

If it is necessary to replace this type of relay, the correct size must be used. The current relay is matched to the characteristics of the motor it is designed to be used with. This type of relay is also sensitive to the position it is mounted in. The current relay generally uses the force of gravity to open the starting contacts. When installing a new relay, it must be mounted in the correct position. If it is installed upside down, the starting contacts will be closed instead of open.

When testing this type of relay, an ohmmeter can be used to check the continuity of the contacts. When the relay is held in the correct position, the ohmmeter should show an open circuit across the contacts. If it does not, the contacts are shorted. If the relay is held upside down, the contacts should indicate continuity. The coil of the relay is generally exposed and a visual inspection will reveal shorted windings. The best method of testing the relay is with an ammeter. If the ammeter is used to measure the current flow to the start winding, it can be seen if the motor starts and the relay contacts disconnect the start winding.

THE POTENTIAL RELAY

The potential (voltage) relay operates by sensing an increase in the voltage developed in the start

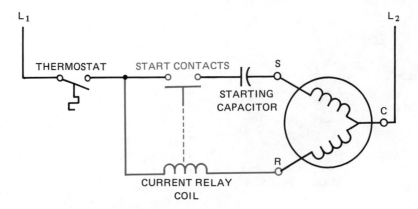

FIGURE 20-4 Current relay connection

FIGURE 20-5 Potential starting relay

winding when the motor is operating. A potential relay is shown in figure 20-5. A schematic diagram for a potential starting relay circuit is shown in figure 20-6. In this circuit, the potential relay is used to disconnect the starting capacitor from the circuit when the motor reaches about 75% of its full speed. SR (starting relay) coil is connected in parallel with the start winding of the motor. A normally closed SR contact is connected in series with the starting capacitor. When the thermostat contact closes, power is applied to both the run and start windings of the motor. Notice that both the run capacitor and the start capacitor are in the circuit at this time.

The rotating magnetic field of the stator cuts

through the bars of the squirrel cage rotor and induces a current into them. The current flow in the rotor produces a magnetic field around the rotor. As the rotor begins to turn, its magnetic field cuts through the start winding and induces a voltage into it. The induced voltage causes the total voltage across the start winding to be higher than the voltage applied by the line. When the motor has accelerated to about 75% of full speed, the induced voltage in the start winding is high enough to energize SR coil. When SR coil energizes, SR contact opens and disconnects the start capacitor from the circuit, figure 20-7.

Notice that this type of relay depends on the voltage produced across the start winding by the applied voltage and the induced voltage of the rotor. For this reason, the potential starting relay is used only to disconnect the starting capacitor from a permanent-split capacitor motor. If this relay was to be used with a motor that disconnected the start winding from the applied voltage, the relay coil would de-energize and SR contact would close. When SR contact closed, it would reconnect the start winding to the circuit.

This type of starting relay is often used with compressors that use the permanent-split capacitor motor with extra starting capacitor. The coil of the relay can be tested for an open circuit with an ohmmeter. When the ohmmeter is connected across the coil, it should indicate continuity. The actual

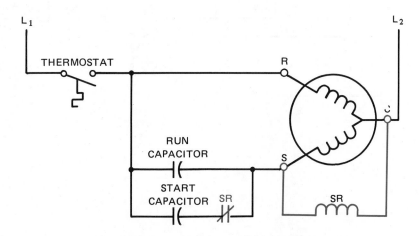

FIGURE 20-6 Potential relay connection

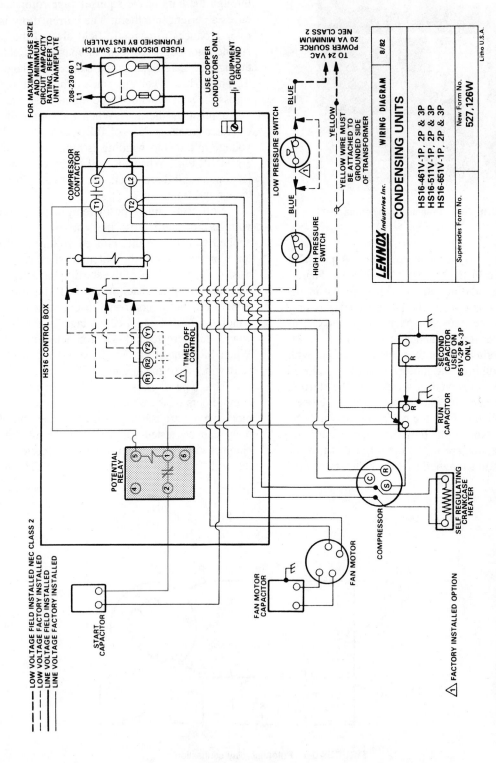

FIGURE 20-7 A potential relay is used to disconnect the start windings of the compressor when the motor reaches about 3/4 of full speed. (Courtesy of Lennox Industries Inc.)

amount of resistance can vary from one type of relay to another. The best method for testing the starting relay is with an ammeter. If an ammeter is connected to the start capacitor, it can be seen if the capacitor is energized when the motor is started, and if the relay disconnects it from the circuit.

SOLID-STATE STARTING RELAYS

Another type of starting relay is known as the solid-state starting relay, figure 20-8. This relay is intended to replace the current-type starting relay, and has several advantages over the current relay. Some of these advantages are

1. The solid-state relay contains no moving parts and no contacts, which can become burned or pitted.
2. The solid-state relay can be used to replace almost any current relay. This interchangeability makes it possible for the service technician to stock only a few solid-state relays instead of a large number of current relays.

The solid-state starting relay is actually an electronic component known as a *thermistor*. A thermistor is a device which exhibits a change of resistance with a change of temperature. This particular thermistor has a positive coefficient of resistance, which means that the resistance of the device will increase with an increase of temperature. The schematic diagram in figure 20-9 illustrates the connection for a solid-state starting relay. Notice that this is the same basic connection used for the connection of a current starting relay, figure 20-4. The solid-state relay, however, does not contain a coil or contacts. When the solid-state relay is used, a current path exists between the line connection terminal and the terminal marked M for MAIN winding. The thermistor is connected between the line connection and the terminal marked S for START winding.

When power is first applied to the circuit, the thermistor has a relatively low resistance. This permits current to flow through both the start and run

FIGURE 20-8 Solid-state starting relay

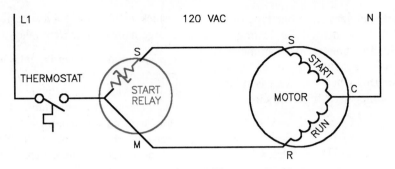

FIGURE 20-9

windings of the motor. The temperature of the thermistor increases because of the current flowing through it. The increase of temperature causes the resistance to change from a very low value of 3 or 4 ohms to several thousand ohms. This increase of resistance is very sudden and has the effect of opening a set of contacts connected in series with the start winding. Although the start winding is never completely disconnected from the power line, the amount of current flow through it is very small, typically 0.03 to 0.05 amps, and does not effect the operation of the motor. This small amount of leakage current maintains the temperature of the thermistor and prevents it from returning to a low resistance. After power has been disconnected from the motor, a cool-down period of about 2 minutes should be allowed before restarting. This cool-down period is needed for the thermistor to return to a low value of resistance.

TESTING THE SOLID-STATE STARTING RELAY

A continuity test can be made with an ohmmeter. If the probes of an ohmmeter are connected to the M and S terminals of the relay, a low value of resistance, typically 2 to 5 ohms, should be seen.

The most accurate test is made by connecting the relay in the motor circuit. A clamp-on ammeter set to its lowest scale can be used to measure the current in the start winding. After the motor has been started, the ammeter should give an indication very close to zero amps.

SOLID-STATE HARD STARTING KIT

Another device which uses a solid-state relay is shown in figure 20-10. This device is intended to increase the starting torque of a permanent-split capacitor motor. The kit contains a solid-state relay and an AC electrolytic capacitor similar to those used as the starting capacitor for a capacitor-start induction run motor. The kit connects directly across the terminals of the existing run capacitor as shown in figure 20-11. When the thermostat contact closes and connects power to the motor circuit, the resistance of the solid-state relay is very low. A current path exists through the run winding, run capacitor, solid-state relay, electrolytic capacitor, and start winding. Since the run capacitor and electrolytic capacitor are connected in parallel, their values of capacitance add, providing extra capacitance to the motor during the starting period. The current flowing through the solid-state relay and electrolytic capaci-

FIGURE 20-10 Hard starting kit increases starting torque of permanent-split capacitor motors. (Courtesy of Motors & Armatures, Inc.)

tor causes the temperature of the relay to increase, resulting in an increase in resistance. The increased resistance reduces the current flow through the electrolytic capacitor to a very low value. This has the effect of disconnecting the electrolytic capacitor from the circuit. The leakage current through the relay and electrolytic capacitor prevents the relay from returning to a low value of resistance. After power has been disconnected from the motor circuit, a cool-down period of 2 to 3 minutes should be given to permit the solid-state relay to return to a low value of resistance.

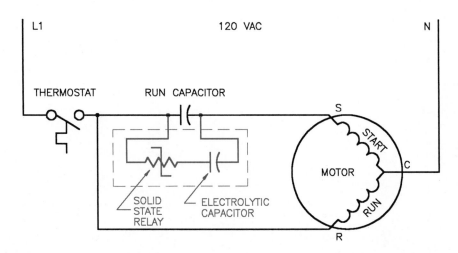

FIGURE 20-11

REVIEW QUESTIONS

1. What are the four types of starting relays?

2. On what type of motor is it necessary to use a starting relay?

3. What principle is used to operate the hot-wire relay?

4. What principle is used to operate the current relay?

5. What type of starting relay does not sense motor current to operate?

6. What type of starting relay can be used for overload protection for the motor?

7. What type of motor can the potential relay be used with?

8. Is the start contact of a hot-wire relay open or closed when power is first applied to the motor?

9. Is the start contact of a current relay open or closed when power is first applied to the motor?

10. Refer to the circuit shown in figure 20-4. What would happen if the coil of the current relay were open when the thermostat connected power to the motor circuit?

UNIT 21
Variable-Speed Motor Control

The use of small variable-speed motors has increased greatly in the last few years. These motors are commonly used to operate light loads such as ceiling fans and blower motors. There are two types of motors used for these applications—the shaded pole and the permanent-split capacitor motor. These motors are used because they operate without having to disconnect a set of start windings with a centrifugal switch or starting relay. Motors intended to be used in this manner are wound with high-impedance stator windings. The high impedance of the stator limits the current flow through the motor when the speed of the rotor is decreased. Speed control for these motors is accomplished by controlling the amount of voltage applied to the motor, or by inserting impedance in series with the stator winding.

VARIABLE VOLTAGE CONTROL

The amount of voltage applied to the motor can be controlled by several methods. One method is to use an autotransformer with several taps, figure 21-1. This type of controller has several steps of speed control. Notice that the applied voltage, 120 volts in this illustration, is connected across the entire transformer winding. When the rotary switch is moved to the first tap, 30 volts is applied to the motor. This produces the lowest motor speed

for this controller. When the rotary switch is moved to the second tap, 60 volts is applied to the motor. This provides an increase in motor speed. When the switch has been moved to the last position, the full 120 volts is applied to the motor operating it at the highest speed.

Another type of variable-voltage control uses a triac to control the amount of voltage applied to the motor, figure 21-2. This type of speed control provides a more linear control since the voltage can be adjusted from 0 to the full applied voltage. At first appearance, many people assume this con-

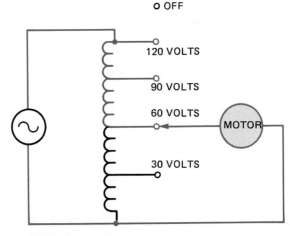

FIGURE 21-1 Autotransformer controls motor voltage.

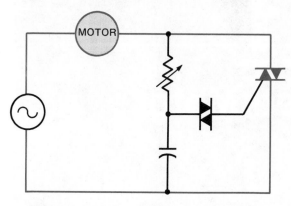

FIGURE 21-2 Triac used to control motor speed

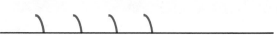

FIGURE 21-3 Triac conducts only the positive half of the waveform.

FIGURE 21-4 Triac speed control

troller to be a variable resistor connected in series with the motor. A variable resistor large enough to control even a small motor would produce several hundred watts of heat and could never be mounted in a switch box. The variable resistor in this circuit is used to control the amount of phase shift for the triac. The triac controls the amount of voltage applied to the motor by turning on at different times during the AC cycle.

A triac speed control is very similar to a triac light dimmer used in many homes. A light dimmer, however, should never be used as a motor speed controller. Triac light dimmers are intended to be used with resistive loads such as incandescent lamps. Light dimmer circuits will sometimes permit one half of the triac to start conducting before the other half. The wave form shown in figure 21-3 illustrates this condition. Notice that only the positive half of the waveform is being conducted to the load. Since only positive voltage is being applied to the load, it is DC. Operating a resistive load, such as an incandescent lamp with DC, will do no damage. Operating an inductive load such as the winding of a motor can do a great deal of damage, however. When direct current is applied to a motor winding, there is no inductive reactance to limit the current. The actual wire resistance of the stator is the only current limiting factor. The motor winding or the controller can easily be destroyed if direct current is applied to the motor. For this reason only triac controllers designed for use with inductive loads should be used for motor control. A photograph of a triac speed controller is shown in figure 21-4.

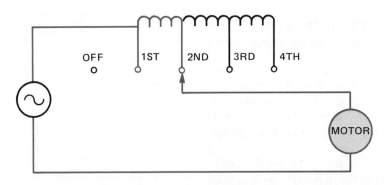

FIGURE 21-5 Series inductor changes impedance of circuit.

SERIES IMPEDANCE CONTROL

Another common method of controlling the speed of small AC motors is to connect impedance in series with the stator winding. This is the same basic method of control used with multi-speed fan motors. The circuit in figure 21-5 shows a tapped inductor connected in series with the motor. When the motor is first started, it is connected directly to the full voltage of the circuit. As the rotary switch is moved from one position to another, steps of inductance are connected in series with the motor. As more inductance is connected in series with the stator, the amount of current flow decreases. This produces a weaker magnetic field in the stator. Rotor slip increases due to the weaker magnetic field and causes the motor speed to decrease. A photograph of this type of controller is shown in figure 21-6.

FIGURE 21-6 Fan speed control using a tapped inductor connected in series with the motor

REVIEW QUESTIONS

1. What two types of small AC motors are used with variable-voltage speed control?

2. Why are these two types of motors used?

3. Name two methods of variable-voltage control for small AC motors.

4. What solid-state device is used to control the voltage applied to the motor?

5. Why is it necessary to use only controllers designed for use with inductive loads?

6. Name a method other than variable voltage used to control the speed of small AC motors.

UNIT 22

The Defrost Timer

Many of the refrigeration appliances used in the home are "frost-free." The frost-free appliance could more accurately be termed "automatic defrost." The brain of the frost-free appliance is the defrost timer. The job of this timer is to disconnect the compressor circuit and connect a resistive heating element located near the evaporator at regular time intervals. The defrost heater is thermostatically controlled and is used to melt any frost formation on the evaporator. The defrost heater is permitted to operate for some length of time before the timer disconnects it from the circuit and permits the compressor to operate again.

TIMER CONSTRUCTION

The defrost timer is operated by a single-phase synchronous motor like those used to operate electric wall clocks, figure 22-1. The contacts are operated by a cam that is gear driven by the clock motor. A schematic drawing of the timer is shown in figure 22-2. Notice that terminal 1 is connected to the common of a single-pole double-throw switch. Terminals 2 and 4 are connected to stationary contacts of the switch. In the normal operating mode, the switch makes connection between contacts 1 and 4. When the defrost cycle is activated, the contact will change position and make connection between terminals 1 and 2. Terminal 3 is connected

to one lead of the motor. The other motor lead is brought outside the case. This permits the timer to be connected in one of two ways, which are:

1. The continuous run timer,
2. The cumulative compressor run timer.

It should be noted that the schematic drawing can be a little misleading. In the schematic shown, the timer contact can only make connection between terminals 1 and 4, or terminals 1 and 2. In actual practice, a common problem with this timer is that the movable contact becomes stuck between terminals 4 and 2. This causes the compressor and defrost heater to operate at the same time.

FIGURE 22-1 Defrost timer

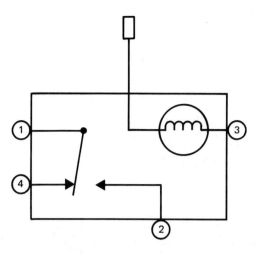

FIGURE 22-2 Schematic of a defrost timer

THE CONTINUOUS RUN TIMER

The schematic for the continuous run timer is shown in figure 22-3. Notice in this circuit that the pigtail lead of the motor has been connected to terminal 1, and that terminal 1 is connected directly to the power source. Terminal 3 is connected directly to the neutral. This places the timer motor directly across the power source, which permits the motor to operate on a continuous basis.

Figure 22-4 shows the operation of the timer in the compressor run cycle. Notice there is a current path through the timer motor and a path through the timer contact to the thermostat. This permits power to be applied to the compressor and evaporator motor when the thermostat closes.

Figure 22-5 shows the operation of the circuit when the timer changes the contact and activates the defrost cycle. Notice there is still a complete circuit through the timer motor. When the timer contact changes position, the circuit to the thermostat is open and the circuit to the defrost heater is closed. The heater can now melt any frost accumulation on the evaporator. At the end of the defrost cycle, the timer contact returns to its normal position and permits the compressor to be operated by the thermostat.

THE CUMULATIVE COMPRESSOR RUN TIMER

This circuit gets its name from the fact that the timer motor is permitted to operate only when the compressor is in operation and the thermostat is closed. The schematic for this circuit is shown in figure 22-6. Notice that the pigtail lead of the clock motor has been connected to terminal 2 instead of terminal 1. Figure 2-7 shows the current path during compressor operation. The timer contact is making connection between terminals 1 and 4. This permits power to be applied to the thermostat. When the thermostat contact closes, current is permitted to flow through the compressor motor, the evaporator fan motor, and the defrost timer motor. In this circuit, the timer motor is connected in series with the defrost heater. The operation of the timer motor is not affected, however, because the impedance of the timer motor is much greater than the resistance of the heater. For this reason almost all the voltage of this circuit is dropped across the timer motor. The impedance of the timer motor also limits the current flow throught the defrost heater to such an extent that it does not become warm.

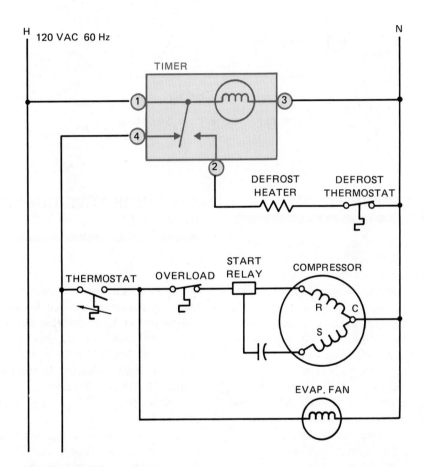

FIGURE 22-3 Schematic of a defrost timer used in a continuous-run circuit

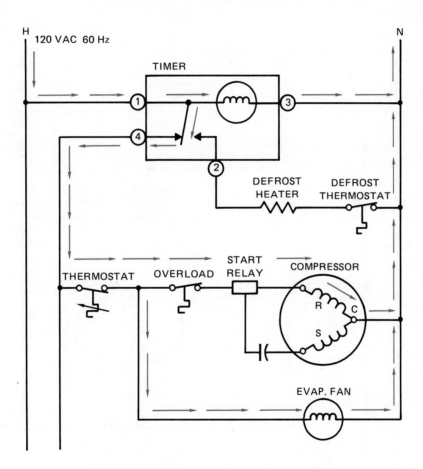

FIGURE 22-4 Current path during cooling operation

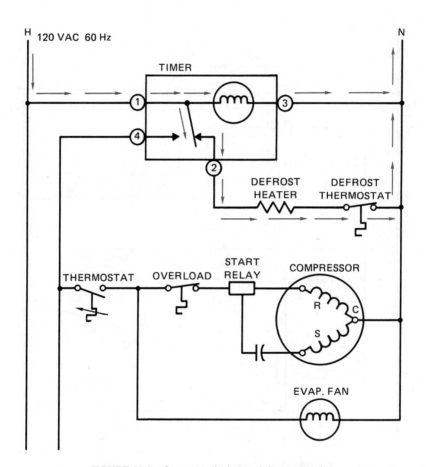

FIGURE 22-5 Current path during defrost operation

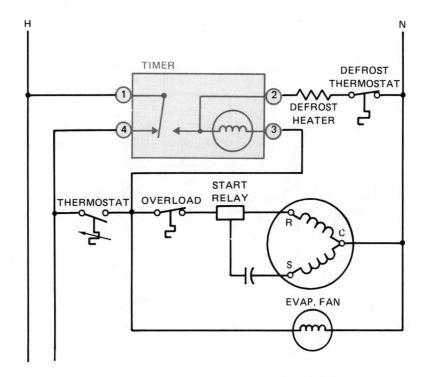

FIGURE 22-6   Defronst timer connected in a comulative compressor run circuit

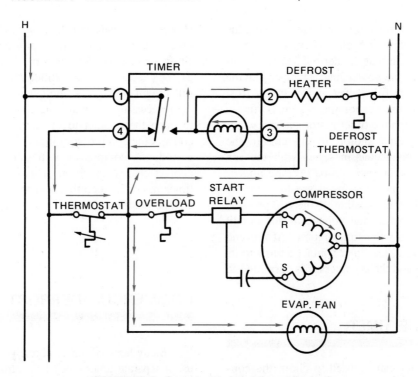

FIGURE 22-7   Current path during the cooling cycle

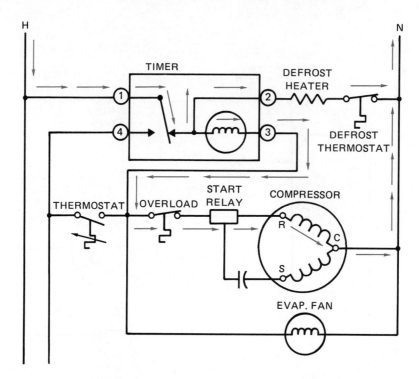

FIGURE 22-8 Current path during the defrost cycle

Figure 22-8 shows the current path through the circuit when the defrost cycle has been activated. Notice in this circuit that the defrost heater is connected directly to the power line. This permits the heater to operate at full power and melt any frost accumulation on the evaporator. There is also a current path through the timer motor and run winding of the compressor motor. In this circuit, the timer motor is connected in series with the run winding of the compressor. As before, the impedance of the timer motor is much greater than the impedance of the run winding of the compressor. This permits almost all the voltage in this circuit to be applied across the timer motor. At the end of the defrost cycle, the timer contact returns to its normal position and the compressor is permitted to operate.

TESTING THE TIMER

An ohmmeter can be used to check the continuity of the contacts and the motor winding.

However, to really test the timer for operation takes time. The cam can be manually turned to the position so that the defrost cycle is turned on. This can be checked with a voltmeter to determine when full circuit voltage is applied to terminal 2. It is then necessary to wait long enough for the timer to open the contact to the defrost heater and reconnect the compressor circuit. If the thermostat is closed, the compressor will start when the timer contact changes position. This test shows that the timer motor is operating and that the contact does change position.

COMMERCIAL DEFROST TIMERS

Many large commercial refrigeration units often use a separate timer clock to control the defrost cycle. This has several advantages over the previously

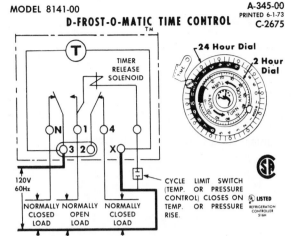

MODEL 8141-00

A-345-00
PRINTED 6-1-73
C-2675

D-FROST-O-MATIC TIME CONTROL

24 Hour Dial

2 Hour Dial

T

TIMER
RELEASE
SOLENOID

N   1   4

3 2   X

120V
60Hz

NORMALLY   NORMALLY   NORMALLY
CLOSED   OPEN   CLOSED
LOAD   LOAD   LOAD

CYCLE LIMIT SWITCH (TEMP. OR PRESSURE CONTROL) CLOSES ON TEMP. OR PRESSURE RISE.

CSA

LISTED
REFRIGERATION
CONTROLLER
518H

INSTALLING AND OPERATING DIRECTIONS:

Place start pins in outer (24 Hour) dial at the time of day that switch contacts are to be reverse from shown above when dial pins are opposite time pointer. **CAUTION:** (Leave at least 1 hole between each adjacent pin).

TO SET BACK-UP DEFROST TERMINATION: Push down and rotate pointer on inside (2 Hour) dial until it is opposite desired time.

TO SET TIME OF DAY: Grasp knob in the center of the inner (2 Hour) dial and rotate it in a counter-clockwise direction. This will revolve the outer dial. Line up the correct time of day on the outer dial with the time pointer. **Do not try to set the time control by grasping the outer dial. Rotate the inner dial only.**

FOR REPLACEMENT of this control contact refrigerator manufacturer.

FOR REPAIR contact nearest Paragon service station.

MAXIMUM CONTACT RATING		
40A	Non Inductive	120Vac
2hp		120Vac
690VA	Pilot Duty	120Vac

Timing Motor: 120V 60Hz

Made in the U.S.A.

606 Parkway Blvd., P.O. Box 28, Two Rivers, WI 54241 U.S.A.

EXPORT SALES OFFICE: Two Rivers, Wisconsin 54241 U.S.A.
Cable: PECO Telex 26-3450 PARAGON TWOR

IN CANADA: PARAGON ELECTRIC P.O. Box 1030 Guelph, Ontario
Division of AMF CANADA LIMITED

Paragon

Printed in U.S.A.

FIGURE 22-9 Schematic diagram of a commercial defrost timer (Courtesy of Paragon Electric Company, Inc.)

day or night the timer turns on. The second setting determines how long the timer is permitted to remain on. The timer shown in this example can be started on even numbered hours of the day or night. The center knob sets how long the contacts are energized before they return to their normal position. Once turned on, the contacts can be set to remain in their energized position for a minimum of two minutes to a maximum of 120 minutes.

This timer has a separate timer release solenoid incorporated into its design. When the timer release solenoid is energized, it causes the contacts to return to their normal de-energized position immediately. This permits the action of some type of external limit switch, such as a temperature or pressure switch, to terminate the defrost cycle.

FIGURE 22-10 Commercial defrost timer (Courtesy of Paragon Electric Company, Inc.)

discussed defrost timer. When this method is used, the timer clock is connected directly across the power line as shown in figure 22-9. This separates the operation of the timer from the operation of the compressor. In this way, the defrost cycle can be started during periods when the unit is in minimum use.

Timers of this type, figure 22-10, generally have two timed settings. One determines the time of

REVIEW QUESTIONS

1. What type of motor is used to operate the timer?

2. Why is one of the motor leads brought outside the timer?

3. Name two ways of connecting the defrost timer.

4. What function does the defrost heater perform?

5. To which terminal is the pigtail lead of the timer motor connected if the timer is to operate continuously?

UNIT 23

The Thermostat

Thermostats are temperature-sensitive switches. They use a variety of methods to sense temperature, and can be found with different contact arrangements. Some thermostats are designed to be used with low-voltage systems, generally 24 volts; and others are designed to be connected directly to line voltage and operate motors and heating units. The advantage of low-voltage thermostats is that they are more economical and safer to use inside the home.

BIMETAL THERMOSTATS

One of the most common methods of sensing temperature is with a bimetal strip. When used as the temperature sensing element of a thermostat, the bimetal strip is generally bent in a spiral that resembles a clock spring. If a contact is attached to the end of the strip and another contact is held stationary, a thermostat is formed, figure 23-1. A small permanent magnet is used to provide a snap action for the contacts.

This type of thermostat is inexpensive and has the advantage of not having to be mounted in a level position. The greatest enemy of an open-contact thermostat is dirt. This is especially true for thermostats designed for low-voltage operation. If poor thermostat contact is suspected, the contacts should be cleaned. This can be done with a strip of hard paper, such as typing paper, and alcohol.

Soak a strip of hard paper in alcohol and place the strip between the contacts. Close the contacts and draw the strip through the closed contacts. This will generally remove any accumulation of dirt and oil. After cleaning, the contacts should be buffed to remove any alcohol residue. This can be done by drawing a piece of dry hard paper through the contacts several times. This type of thermostat is shown in figure 23-2.

MERCURY CONTACT THERMOSTAT

Another type of contact used with the bimetal type of thermostat is the mercury contact. In this type of thermostat, a small pool of mercury is sealed

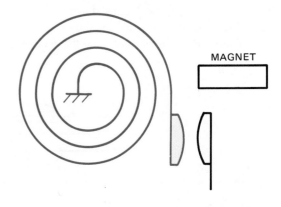

FIGURE 23-1   Contacts operated by a bimetal strip

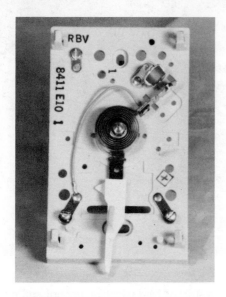

FIGURE 23-2 Open-contact thermostat

inside a glass container. A set of contacts is also sealed in the glass. Most mercury-type contacts are made to be single-pole double-throw, which means there is a common terminal, a normally open terminal, and a normally closed terminal. Figure 23-3 illustrates this type of contact. Notice in this example that the pool of mercury makes connection with the common terminal, located in the center, and the normally closed terminal. If the glass bulb is tilted in the other direction, the mercury will flow to the opposite end and make connection between the common terminal and the normally open contact.

The mercury contact has the advantage of being sealed in glass and not subjected to dirt and oil. When this type of contact is used with a bimetal strip, it is generally mounted as shown in figure 23-4. This type of thermostat uses the weight of

the mercury to provide a snap action for the contact instead of a magnet. When the bimetal strip has turned far enough to permit the mercury to flow from one end of the glass bulb to the other, the weight of the mercury prevents any spring action of the bimetal strip from snapping the contact open. A mercury thermostat, however, must be mounted in a level position if it is to operate properly. This can sometimes be a problem in homes that do not remain level. A mercury type thermostat is shown in figure 23-5.

HEATING AND COOLING THERMOSTATS

Many thermostats are designed to be used for both heating and cooling applications. This can be done with thermostats that contain both a normally open and normally closed contact. A simple schematic diagram of this type of thermostat is shown in figure 23-6. Notice the thermostat contact is a single-pole double-throw type. The selector switch is a double-pole type. The dashed line indicates mechanical intertie. With the selector switch in the position shown, the thermostat is being used for heating. If the selector switch is changed, the bottom movable contact will break connection with its stationary contact, and the top movable contact will make connection with its stationary contact. Notice that changing this switch will also change the sense of the thermostat. In the heating position, the thermostat activates the heating unit when the contact closes because of a decrease in temperature. In the cooling position, the thermostat activates the air-conditioning unit when the contact makes connection because of an increase in temperature.

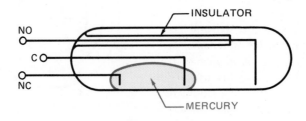

FIGURE 23-3 Mercury contacts

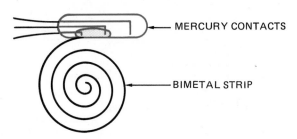

FIGURE 23-4   Mercury-type thermostat

THE FAN SWITCH

Many thermostats are designed to permit manual control of the blower fan. This is done to permit the blower fan to be operated separately. Some people find it desirable to operate the blower fan continuously to provide circulation of air throughout the building. This is especially true for buildings equipped with electronic air cleaners (precipitators) or for buildings that must remove undesirable elements such as smoke in an office building or night club. A schematic diagram of this type of circuit is shown in figure 23-7. The fan switch is a single-pole double-throw switch. When the switch is in one position, it permits the fan relay to be controlled by the thermostat. If the switch is thrown

in the opposite direction, the fan relay is connected directly to the control voltage.

THE HEAT ANTICIPATOR

The heat anticipator is a small resistance heater located near the bimetal strip. The function of this heater is to slightly preheat the bimetal strip and prevent overrun of the heating system. For example, many heating systems, such as fuel oil or gas, operate by heating a metal container called a heat exchanger. When the temperature of the heat exchanger reaches a high enough level, a thermostatically controlled switch causes the blower to turn on and draw air across the heat exchanger. The moving air causes heat to be removed from the heat exchanger to the living area. When the thermostat is satisfied, the heating unit is turned off. The blower will continue to operate, however, until the excess heat has been removed from the heat exchanger.

Now assume that the thermostat has been set for a temperature of 75 degrees. If the heating unit is permitted to operate until the temperature reaches 75 degrees, the final temperature of the living area may be from 3 to 5 degrees higher than the thermostat setting by the time enough heat has been removed from the heat exchanger to cause the blower to turn off.

If the heat anticipator has been properly set, however, it will cause the thermostat to turn off several degrees before the room temperature has reached the thermostat setting. This permits the excess heat of the heat exchanger to raise the temperature to the desired level without overrunning the thermostat setting.

The setting of the heat anticipator is controlled

FIGURE 23-5   Mercury thermostat

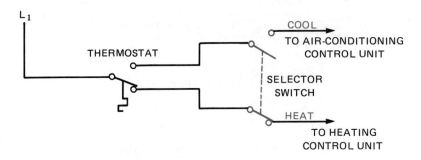

FIGURE 23-6 Dual operation of a thermostat

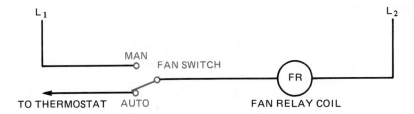

FIGURE 23-7 Fan switch

by a sliding contact. There are markings such as .2, .25, .3, .35, and .4. The sliding contact is generally set at the number that corresponds to the current rating of the control system. The current rating can generally be located in the service information or on the control unit itself. The heat anticipator does not have to be set at that position, however. The service technician should set it to operate the unit for longer or shorter periods depending on the desires of the customer.

THE COOLING ANTICIPATOR

A device which operates in a similar manner to the heat anticipator is the cooling anticipator. The cooling anticipator is a resistive heating element which operates in an opposite sense to the heat anticipator. The cooling anticipator operates while the thermostat contacts are open and the air conditioning unit is not running. The cooling anticipator heats the thermostat slightly and causes it to close its contacts before the ambient temperature increases enough to close them.

The circuit shown in figure 23-8 displays the current path of the heat anticipator for a heating and cooling thermostat during the heating cycle. In this mode of operation, current flows through the heat anticipator while the thermostat contact is closed and the heating unit is in operation.

In figure 23-9, the thermostat has been switched to the cooling mode. Notice that a current path exists through the cooling anticipator when the thermostat contact is open and the air conditioning unit is not in operation. When the thermostat contact closes, a low resistance path exists around the cooling anticipator. This stops the flow of current through it while the air conditioning unit is in operation.

LINE VOLTAGE THERMOSTATS

Line voltage thermostats are generally used to control loads such as blower fans and heating elements. This means that the thermostat must contain contacts that are capable of handling the current needed to operate these loads without an intervening relay. This type of thermostat is shown in fig-

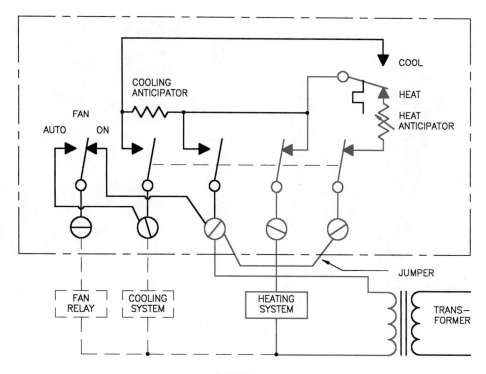

FIGURE 23-8

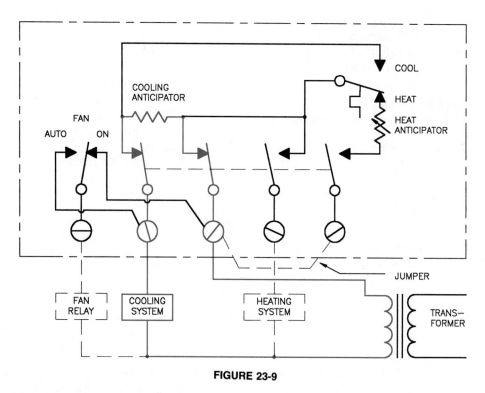

FIGURE 23-9

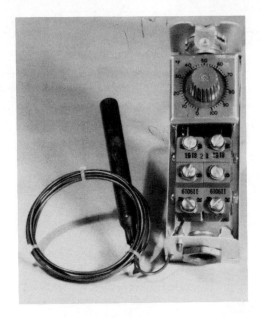

FIGURE 23-10 Line voltage thermostat

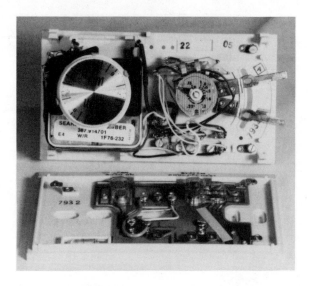

FIGURE 23-12 Programmable thermostat

ure 23-10. Many of these thermostats use the pressure of refrigerant in a sealed system to sense temperature, figure 23-11. When the temperature increases, the pressure in the system increases also. The increase in pressure causes the bellows to expand. When the bellows expands far enough, it activates a set of spring-loaded contacts.

PROGRAMMABLE THERMOSTATS

The term "programmable" is a catch word that has taken on many meanings. In the case of ther-

mostats, the term programmable generally means a thermostat that can be set to operate at different temperature settings at different times. They range in complexity from units that use a simple time clock to units that are operated by integrated circuits (ICs), and permit the temperature to be set to any desired level at any desired time. A programmable thermostat is shown in figure 23-12. This thermostat uses a quartz-operated time clock and two separate thermostat units. The time clock is used to operate a switch. The setting of the clock determines the position of the switch at any particular time. A schematic for this type of circuit is shown in figure 23-13. Notice that the position of the clock-oper-

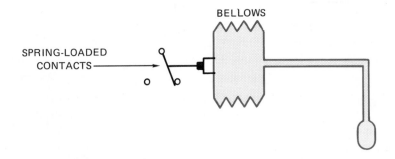

FIGURE 23-11 Thermostat contacts are operated by pressure.

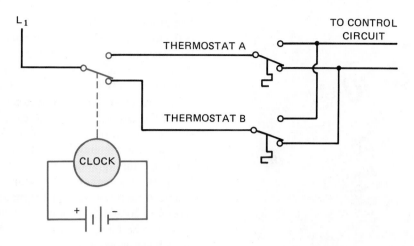

FIGURE 23-13 Schematic for a programmable thermostat

ated switch determines which thermostat is used to control the system. To understand the operation of this system, assume that thermostat "A" has been set for a temperature of 95 degrees, and thermostat "B" is set for 75 degrees. The time clock has been set to permit thermostat "A" to control the air-conditioning system when there is no one in the residence. One hour before people are to return home, the time clock changes the contact and thermostat "B" is used to control the system. Since thermostat "B" has been set for 75 degrees, the residence will have been cooled to that temperature when the people arrive.

The programmable thermostat can reduce energy consumption by maintaining a desired temperature only during the hours the dwelling is occupied. The temperature can be maintained at an uncomfortable level the rest of the time, which permits the air-conditioning unit to operate much less.

STAGING THERMOSTATS

Staging thermostats are similar to programmable thermostats in that they contain two separate sets of contacts. Unlike the programmable thermostat, however, the staging thermostat contains only one bimetal strip, which is used to control the action of both sets of contacts. One set of contacts is designed to operate slightly behind the other set. A good example of how a staging thermostat is used can be found in a heat-pump system. Assume that the first contact is used to operate the compressor relay, and the second contact is used to operate the contactor, which controls the electric resistance heating strips. When the temperature decreases, the first thermostat contact closes and connects the compressor to the line. If the compressor is able to provide enough heat to the dwelling, the second contact will never make connection. If, however, the outside temperature is low enough that the heat pump cannot provide enough heat, the second thermostat contact will close and permit the electric heat strip to operate.

THE DIFFERENTIAL THERMOSTAT

The differential thermostat is used primarily with solar-powered heating systems. A differential thermostat is shown in figure 23-14. This thermostat uses two separate temperature sensors and is activated by the difference of temperature between them. A solar hot-water system is shown in

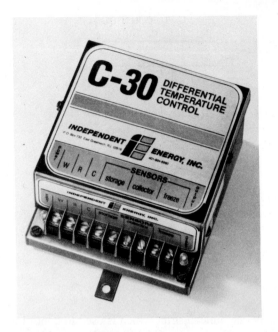

FIGURE 23-14 Differential thermostat (Courtesy of Independent Energy Corp.)

ferential thermostat. When the temperature of the collector becomes greater than the temperature of the water in the storage tank by so many degrees, the thermostat activates the pump motor. The pump motor circulates water from the storage tank to the collector, and from the collector back to the tank. When the temperature of the collector is within a certain amount of the water temperature, the thermostat turns the pump off. In this way, water is circulated through the collector only when the collector is at a higher temperature than the stored water. A common setting for the differential thermostat is 20 and 5. This means that the thermostat will turn the pump on when the collector is 20 degrees hotter than the stored water, and turn the pump off when the collector is only 5 degrees hotter than the stored water.

Some differential thermostats provide extra features, such as antifreeze protection. Antifreeze protection turns the pump on and circulates warm water through the collector when its temperature is near freezing. This does cool the warmed water, but this is generally preferred to damaging the collector. Some solar systems used a separate water supply for the collector. These systems use a mixture of antifreeze and water in the collector loop to avoid freezing problems.

figure 23-15. A solar collector is used to heat the water. A storage tank stores the heated water and acts as a heat exchanger for the domestic hot water for the home. The system is controlled by the dif-

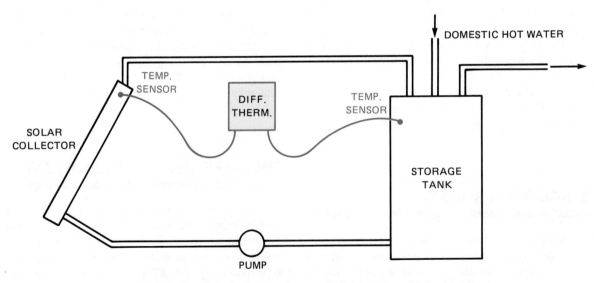

FIGURE 23-15 Solar hot-water system controlled by differential thermostat

SECTION 4 CONTROL COMPONENTS

REVIEW QUESTIONS

1. What is a thermostat?

2. What is the advantage of an open-contact thermostat?

3. What is the disadvantage of an open-contact thermostat?

4. What is the advantage of a mercury thermostat?

5. What is used to provide a snap action for the contacts in an open-contact type of thermostat?

6. What is used to provide a snap action for the mercury thermostat?

7. What method of sensing temperature is often used with line voltage thermostats?

8. What is a programmable thermostat?

9. What is the advantage of the programmable thermostat?

10. What is a differential thermostat?

11. What are differential thermostats generally used to control?

12. What is antifreeze protection in reference to a differential thermostat?

13. What is the advantage of a low-voltage thermostat over a line voltage thermostat?

14. What is the purpose of the heat anticipator?

15. How is the setting of the heat anticipator generally determined?

UNIT 24

High- and Low-Pressure Switches

High- and low-pressure switches are used to sense the amount of pressure in an air-conditioning and refrigeration system. They are used to disconnect the compressor from the power line if the pressure should become too high or too low. Most of the pressure switches used for air conditioning are operated by a bellows. A tube is attached to one end of the bellows and the other end is connected to the discharge or suction side of the compressor, depending on which type of pressure switch is used.

THE HIGH-PRESSURE SWITCH

Figure 24-1 illustrates the operation of a high-pressure switch. The bellows is connected to the discharge side of the compressor via the tube. As the pressure of the system increases, the bellows expands. The bellows is used to activate a spring-loaded normally closed switch. If the pressure should become too great, the bellows will expand far enough to open the switch. The normally closed pressure switch is connected in series with the compressor circuit shown in figure 24-2. The pressure switch may be connected in series with the compressor, or in series with the compressor control relay, depending on the type of control circuit.

THE LOW-PRESSURE SWITCH

The low-pressure switch is very similar in construction to the high-pressure switch. The low-

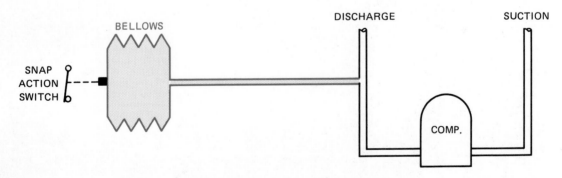

FIGURE 24-1 Pressure switch connected to sense high pressure

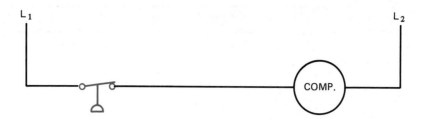

FIGURE 24-2 The pressure switch opens when the pressure becomes too high.

FIGURE 24-3 The pressure switch opens when the pressure switch becomes too low.

pressure switch, however, is connected to the low-pressure or suction side of the compressor. The low-pressure switch is used to disconnect the compressor from the circuit if the pressure on the suction side should become too low. Figure 24-3 illustrates this type of circuit. The low-pressure switch is a normally open held-closed switch. The switch is held in the closed position by the pressure of the system. If the pressure should drop low enough, the switch will open and disconnect the compressor from the circuit. As with the high-pressure switch, the low-pressure switch can be used to disconnect the compressor from the line or disconnect the compressor relay, depending on the type of control circuit.

CONSTRUCTION

Most of the pressure switches used for commercial or industrial systems are adjustable, figure 24-4. This feature allows the service technician to

FIGURE 24-4 Adjustable pressure switch (Courtesy of Johnson Controls Inc.)

PRESSURE SWITCH SETTINGS				
Type of Refrigerant	High Pressure		Low Pressure	
	Cut out	Cut in	Cut out	Cut in
12	225	145	15	35
22	380	300	38	68
500	280	200	22	46

FIGURE 24-5

use the switch on different systems. For example, the pressure settings are different for different types of refrigerant. Figure 24-5 shows a table of common pressure settings for high- and low-pressure switches used with different types of refrigerant.

A dual-pressure switch is shown in figure 24-6. This switch incorporates both high- and low-pressure switches in the same housing. Some manufacturers use a nonadjustable type of pressure switch. This switch is commonly found on central units designed for residential use. This type of switch is shown in figure 24-7. The advantage of this type of switch is that it is inexpensive. When replacing this type of pressure switch, however, it must be matched to the refrigerant system. High- and low-pressure switches are shown in the schematic in figure 24-8.

FIGURE 24-6 Dual pressure switch (Courtesy of Johnson Controls Inc.)

FIGURE 24-7 Non-adjustable pressure switch

SECTION 4 CONTROL COMPONENTS

LEGEND	
CMT	CONDENSER MOTOR THERMOSTAT
CAP	CAPACITOR
CC	COMPRESSOR CONTACTOR
CCH	CRANKCASE HEATER
CFM	CONDENSER FAN MOTOR
COMP	COMPRESSOR
EFC	EVAPORATOR FAN CONTACTOR
F	FUSE
RR	RESET RELAY
T	TRANSFORMER
TB	TERMINAL BLOCK
TOP	THERMAL OVERLOAD PROTECTOR
TS	TERMINAL STRIP
EFM	EVAPORATOR FAN MOTOR
LPC	LOW PRESSURE CONTROL
HPC	HIGH PRESSURE CONTROL
OL	OVERLOAD PROTECTOR

— WARNING —
DISCONNECT ELECTRICAL POWER
SOURCE TO PREVENT INJURY OR
DEATH FROM ELECTRICAL SHOCK

— CAUTION —
USE COPPER CONDUCTORS ONLY
TO PREVENT EQUIPMENT DAMAGE

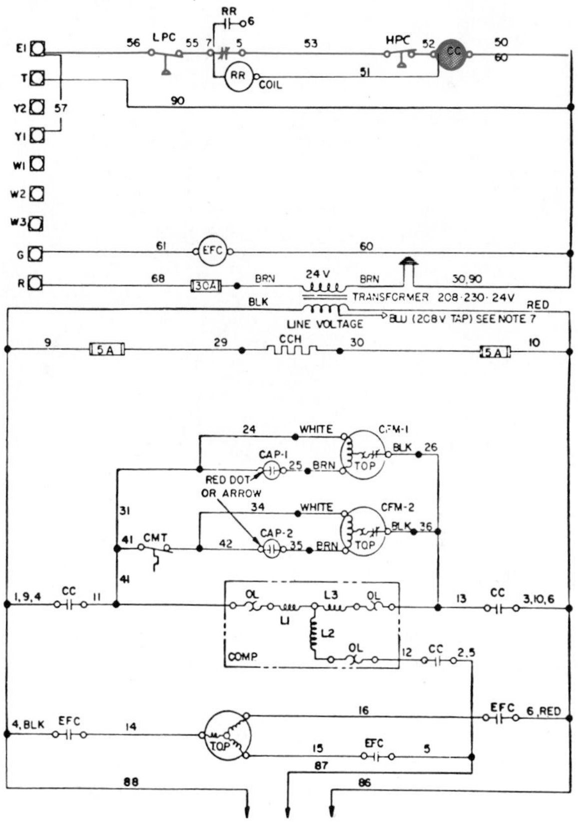

FIGURE 24-8 High- and low-pressure switches are used to disconnect the compressor contactor (Courtesy of Trane Corp.)

REVIEW QUESTIONS

1. What device is used to construct most of the pressure switches used in the air-conditioning field?

2. What type of contact is used with a high-pressure switch?

3. What type of contact is used with a low-pressure switch?

4. Where in the refrigerant system is the high-pressure switch connected?

5. Where in the refrigerant system is the low-pressure switch connected?

UNIT 25

The Flow Switch

The type of flow switch used in air-conditioning systems senses the flow of air instead of the flow of liquid. The flow switch is often referred to as a *sail switch* because it operates on the principle of a sail. The flow switch is constructed from a snap-action microswitch. A metal arm is attached to the microswitch. A piece of thin metal or plastic is connected to the metal arm. The thin piece of metal or plastic has a large surface area and offers resistance to the flow of air. When a large amount of airflow passes across the sail, enough force is produced to cause the metal arm to operate the contacts of the switch. A flow switch is shown in figure 25-1.

The flow switch is used to give a positive indication that the evaporator or condenser fan is operating before the compressor is permitted to operate. The airflow switch is the only positive method of indicating that the fan is actually in operation. For example, in the circuit shown in figure 25-2, the thermostat controls the operation of a control relay. When the thermostat closes its contacts, CR coil energizes. This causes all CR contacts to close. When the first CR contact closes, CFM (Condenser Fan Motor) relay energizes and starts the condenser fan motor. When the second CR contact closes, EFM (Evaporator Fan Motor) relay coil energizes and starts the evaporator fan motor. The third CR contact cannot energize the compressor relay coil, however, because it is interlocked with CFM and EFM relays. The compressor relay coil

can be energized only after the condenser fan and evaporator fan relay coils have energized.

The idea behind this type of control is to ensure that the compressor cannot be started until both the condenser and evaporator fans are operating. This control circuit, however, does not fulfill that requirement. This circuit does not sense if the fans are actually operating. It does sense if the relay coils, which control those fan motors, are ener-

FIGURE 25-1 Airflow switch (Courtesy of Honeywell Inc.)

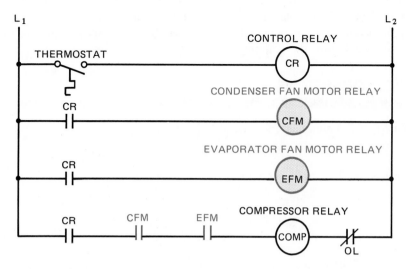

FIGURE 25-2 Compressor is interlocked with condenser and evaporator fan relays.

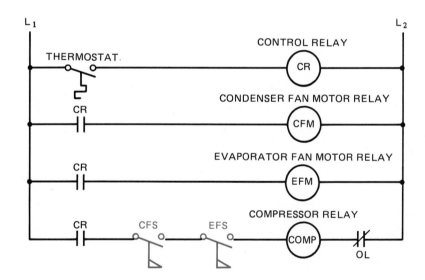

FIGURE 25-3 Compressor is interlocked with airflow switches.

gized. This circuit cannot detect if a fan motor is not operating, or if a belt is broken between the motor and the fan.

The circuit shown in figure 25-3 has been modified from the circuit in figure 25-2. Notice in this circuit that the normally open CFM and EFM contacts connected in series with the compressor relay have been replaced with airflow switches CFS (Condenser Flow Switch) and EFS (Evaporator Flow Switch). These switches are operated by the force of air created by the condenser fan or the evaporator fan. In this circuit, the compressor can be started only after the condenser and evaporator fans are actually operating. If the circuit is in operation and one of the fans should stop, the compressor relay will be disconnected from the circuit. This will disconnect the compressor motor from the circuit.

REVIEW QUESTIONS

1. What is a common name for the airflow switch?

2. What function does the airflow switch perform in a circuit?

3. What is interlocking in a control circuit?

UNIT 26
The Humidistat

The control of humidity can be very important in some heating and air-conditioning systems. Some industries, such as mills that knit polyester and nylon fibers, must maintain a constant humidity because these materials contract and expand with a change of humidity. The control of humidity is also important in heating systems. The amount of humidity in the air has a great effect on the comfort of the living area. If the humidity is to be maintained at a constant level, some device must be used to detect the amount of humidity and then operate some type of control.

The humidistat is a device that can sense the amount of humidity in the air and activate a set of contacts if the humidity should become too high or too low. The two most common materials used to sense humidity are hair and nylon. The materials contract and expand with a change in the amount of humidity in the air. A humidistat using hair as the sense element is shown in figure 26-1.

If a humidifier is used in a central-heating or air-conditioning system, it is generally operated only when the blower is in operation. For this reason, some means is used to interlock the humidifier with the blower. The circuit shown in figure 26-2 uses a humidistat to control the operation of a solenoid coil. The solenoid coil operates a valve that supplies water to the humidifier. Notice that the solenoid coil is interlocked with an airflow switch. The coil can be energized only when the sail switch indicates there is airflow in the system. Some controls use a combination of a humidistat and a sail switch. Nylon strips are used to sense the amount of humidity in the air and a plastic sail is used to sense the flow of air in the duct.

FIGURE 26-1 Humidistat (Courtesy of Honeywell Inc.)

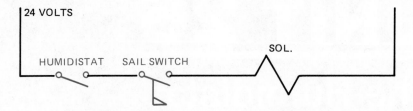

FIGURE 26-2 Humidifier is interlocked with an airflow switch.

REVIEW QUESTIONS

1. What is a humidistat?

2. What are the two most common materials used to sense humidity?

3. What type of control is often used to interlock the humidifier with the blower?

UNIT 27
Fan-Limit Switches

The blower fan of a heating system is generally not permitted to operate until the heat exchanger reaches a high enough temperature to ensure that cold air will not be delivered into the living area. Fan switches are generally operated by a bimetal strip that closes a set of contacts when the temperature of the heat exchanger reaches a high enough level. A fan switch of this type is shown in figure 27-1. The control shown on the face of the switch is used to determine the temperature at which the fan will turn off. The temperature at which the switch contacts will close is determined by the manufacturer. Longer operation time of the fan will generally increase the overall efficiency of the heating unit because more of the heat is delivered to the living area and less escapes to unheated areas. Some people, however, do not like the cooler air being delivered by the blower at the end of a heating cycle. For this reason, the switch can be set to turn off sooner and prevent this problem.

Some fan switches are designed with large enough contacts to permit them to control the operation of the blower motor without a fan relay. Other fan switches have small auxiliary contacts and are used to control the coil of a fan relay. The circuit in figure 27-2 shows a fan switch being used to control a blower motor. Notice the schematic symbol is the same as a thermostat. This symbol is used because it is a thermally-activated switch.

Another type of fan switch is shown in figure 27-3. This type of switch does not sense the heat

of a heat exchanger to close a set of contacts. This switch is basically a timer. It uses a small resistance heater which is controlled by the thermostat. The heater causes a bimetal strip to bend. When the strip has bent far enough, it closes a set of spring-loaded contacts and connects the motor to the line. A schematic of this relay is shown in figure 27-4. Notice that this switch has two sections that are isolated from each other. The 24-volt section is connected to the heating element. The 120-volt section is connected to the switch contacts. When

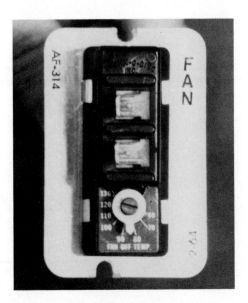

FIGURE 27-1 Fan switch

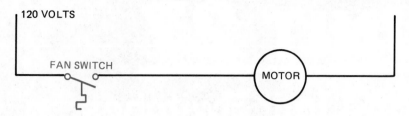

FIGURE 27-2 Fan switch controls the operation of the motor.

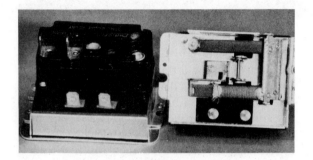

FIGURE 27-3 Time-delay fan switch

connecting this type of fan switch, care must be taken not to connect the terminals to the wrong voltage. The advantage of this type of switch is that it can be used to replace almost any type of thermally-operated fan switch because it can be mounted almost anywhere.

Another type of fan switch is shown in figure 27-5. This switch is used to control the speed of a fan motor as opposed to turning it on or off. This switch is more common to an air-conditioning system than a heating system. It is used to control the speed of the condenser fan. Some systems decrease the speed of the condenser fan if the temperature of the condenser drops too low. A low condenser temperature will cause a low head pressure. This switch is basically a single-pole double-throw switch, figure 27-6. When the condenser temperature is low, the motor is connected to low-speed operation. When the condenser temperature increases, the motor is connected for high-speed operation.

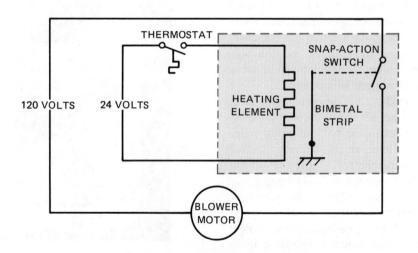

FIGURE 27-4 Timer used as a fan switch

FIGURE 27-5   Bimetal fan-speed switch

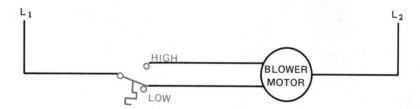

FIGURE 27-6   Condenser speed-control switch

LIMIT SWITCHES

One type of limit switch is used as a safety cut-off switch for a heating system. Limit switches are generally a bimetal-operated switch. Two types of limit switches are shown in figure 27-7. Limit switches contain a normally closed contact, which is connected in series with the system control. They are not, however, connected in series with the blower fan, figure 27-8. Notice tht the blower motor can operate independently of the burner control. This permits the blower to cool down the heating unit if the high-limit switch should open and turn the burner off.

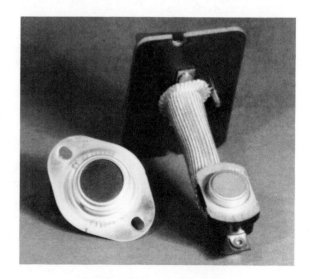

FIGURE 27-7   High-limit switches

UNIT 27   FAN-LIMIT SWITCHES

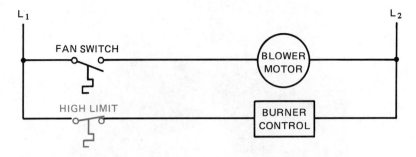

FIGURE 27-8 Limit switch is connected in series with the burner control.

FIGURE 27-9 Fan-limit switch

FAN-LIMIT SWITCH

The fan-limit switch contains both a fan switch and a high-limit switch in one housing, figure 27-9. This type of switch uses a bimetal strip formed in the shape of a spiral. When the bimetal is heated, it causes a cam to rotate. When the cam has turned far enough, the normally open fan switch closes and connects the blower fan to the line, figure 27-10. If the system is operating properly, the blower fan prevents the temperature of the heat exchanger from rising high enough to open the high-limit switch contacts. If the blower fan does not operate, however, the temperature of the heat exchanger will increase enough to open the normally closed limit switch. This will cause the burner to turn off and prevent damage to the heating system.

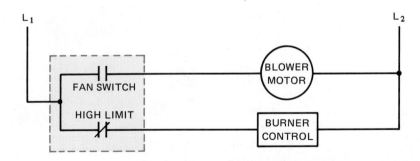

FIGURE 27-10 Schematic for fan-limit switch

REVIEW QUESTIONS

1. Why do some fan switches permit the temperature at which the switch will turn off to be set?

2. What type of sensing device do most fan switches use to determine when the temperature is high enough to start the blower fan?

3. What type of contact arrangement is used for switches that control the speed of a condenser fan motor?

4. What is the most common use for a high-limit switch?

5. Why is the blower fan not connected in series with the limit switch?

UNIT 28

The Oil-Pressure Failure Switch

Many of the larger air-conditioning units use a forced-oil system for the compressor instead of a splash system. When a forced-oil system is used, an oil pressure failure switch is often employed to protect the compressor from insufficient oil pressure. The oil-pressure failure switch actually contains several control functions in the same unit. These functions include a differential pressure switch, a timer, and a set of control contacts.

THE DIFFERENTIAL PRESSURE SWITCH

The actual oil pressure in a compressor is the difference in pressure between the suction pressure and the discharge pressure of the oil pump. For example, if the suction pressure is 35 PSI, and the oil pump discharge pressure is 65 PSI, the actual amount of oil pressure in the system is 30 PSI (65 − 35 = 30). If the oil-pressure failure switch is to be used to measure the actual oil pressure in the system, it must be able to measure the difference between these two pressures. This is accomplished with a differential pressure switch. Figure 28-1 illustrates the operation of an oil-pressure failure switch. Notice in this illustration that two bellows are employed. One bellows is connected to the suction side of the compressor. The other bellows is connected to the oil pump discharge. If the oil pressure is low, the bellows connected to the suc-

tion side of the compressor forces the differential pressure switch to remain closed. If the oil pressure increases high enough above the pressure of the suction line, the oil pressure bellows will provide enough force to overcome the force of the suction line bellows and open the differential pressure switch. Notice that the differential pressure switch remains closed until there is enough oil pressure to open it.

THE TIME-DELAY CIRCUIT

The differential pressure switch is used to control the time-delay circuit. The time-delay circuit consists of a current-limiting resistor, a resistance heating element, and a bimetal strip. Most oil pressure failure switches are designed to be used on 120- or 240-volt connections. This selection is made possible by the value of the current-limiting resistor. Notice that this resistor is center tapped. If 240 volts is to be applied to the circuit, the full value of the current-limiting resistor is connected in series with the heater. If the circuit is 120 volts, the line is connected to the center tap position of the current-limiting resistor. Since the value of the resistor is cut in half for the 120-volt connection, the same amount of current will flow through the heater for either line voltage.

The resistance heater is used to heat the bimetal strip. If the heater is permitted to operate long

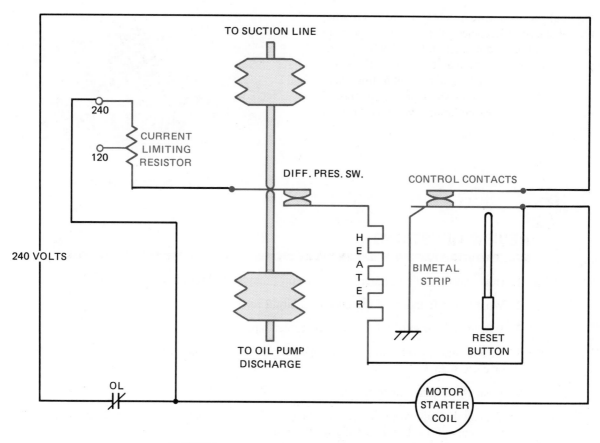

FIGURE 28-1 Schematic of oil pressure failure switch

enough, the bimetal strip will warp away, and the control contact will open. The time-delay circuit is necessary to permit the compressor to operate long enough for oil pressure to build up in the system. When the oil pressure reaches a high enough level, the differential pressure switch opens and disconnects the heater from the circuit. This stops the warping action of the bimetal strip and the control contacts do not open.

THE CONTROL CONTACTS

Notice that the control contacts are connected in series with the motor starter coil to the compressor. If the control contacts should open, the circuit to the motor starter will be broken and the

compressor will be disconnected from the line. Notice that the control contacts provide power to the heater of the timer. If the control contacts should open, power cannot be applied to the heater circuit until the contacts are closed. Once the contacts have opened, they must be manually reset by the reset button.

SPECIFICATIONS

The normal usable oil pressure for most reciprocating compressors is generally between 35 and 45 PSI. The differential pressure control permits the cut-in and cut-out points to be set. A common setting for this type of switch is cut-in at 18 PSI and cut-out at 12 PSI. This means that the differential pressure switch contacts will open when the oil pressure becomes 18 PSI greater than the pressure of the suction line, and close when the oil pressure drops to a point that it is only 12 PSI above the suction line pressure. The amount of time delay is set by the manufacturer and is generally about 2 minutes. An oil-pressure failure switch is shown in figure 28-2.

FIGURE 28-2 Oil pressure failure switch

REVIEW QUESTIONS

1. How can the actual amount of useful oil pressure in a compressor be found?

2. What is the function of the current-limiting resistor?

3. Why is the current-limiting resistor center tapped?

4. Does a high enough oil pressure open the differential pressure switch contacts or close them?

5. What is the function of the heater?

6. Explain the sequence of events that take place if the oil pressure does not become great enough to disconnect the heater circuit.

7. What is the cut-in point?

8. What is the cut-out point?

9. Is the timer circuit connected in series with the motor starter coil?

10. Are the control contacts connected in series with the motor starter coil?

UNIT 29
Solenoid Valves

A solenoid valve is an electrically-operated valve. They are used to control the flow of gasses or liquids. They range in complexity from a simple on-off valve to 4-way reversing valves used on heat-pump systems. A simple plunger-type of solenoid valve is shown in figure 29-1. This type of valve is often used to control the flow of gas or liquid in an air-conditioning system. The plunger of the solenoid is used to lift the valve off its seat. The valve is held closed by a spring when it is in its normal or de-energized position. When the coil is energized, the plunger lifts the valve off the seat and liquid or gas is permitted to flow from the inlet to the outlet. When the coil is de-energized, the spring returns the valve to the seat and stops the flow of liquid or gas.

Notice that the valve is marked with an inlet and outlet side. The inlet is connected to the side of the system with the highest pressure. In this way, the pressure of the system is used to help keep the valve closed. If the valve should be reversed and pressure applied to the outlet side, the pressure of the system could be enough to overcome the tension of the spring and lift the plunger off the seat. This would cause the valve to leak.

THE REVERSING VALVE

A very common solenoid valve used in the air-conditioning field is the 4-way valve or reversing valve. Reversing valves are used to change the direction of flow of refrigerant in a heat-pump system. Figure 29-2 shows the direction of refrigerant flow when the heat-pump unit is in the cooling cycle. Notice that the high-pressure gas leaving the compressor enters the reversing valve. It is then directed to the outside coil being used as the condenser during the cooling cycle. Liquid refrigerant flows from the outside coil to the metering device where it is changed to a low-pressure liquid. The low-pressure liquid then enters the inside coil where it attracts heat from the inside air and changes to a gas. It then flows to the reversing valve. The reversing valve directs the flow to the accumulator and back to the compressor.

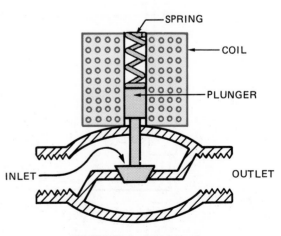

FIGURE 29-1 Solenoid valve

169

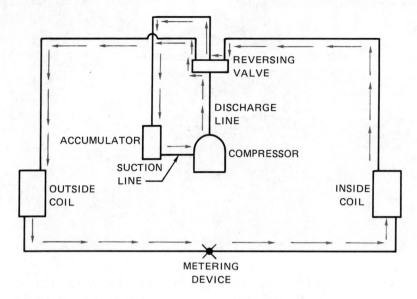

FIGURE 29-2 Refrigerant flow during the cooling cycle

If the unit is now to be used for heating, the flow of refrigerant must be reversed through the system, figure 29-3. Notice that the flow of hot, high-pressure gas is still from the discharge side of the compressor to the reversing valve. In this example, however, the flow of high-pressure gas is directed to the inside coil, which is now being used as the condenser. Liquid refrigerant leaves the in-side coil and flows to the metering valve. The refrigerant is changed into a low-pressure liquid after going through the metering valve and flowing to the outside coil. Heat is then added to the liquid from the surrounding outside air. The gas then flows to the reversing valve, the accumulator, and back to the suction line of the compressor.

Notice in both examples that the direction of

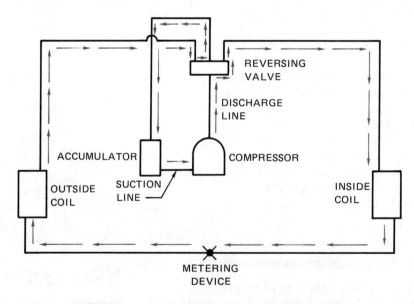

FIGURE 29-3 Refrigerant flow during the heating cycle

SECTION 4 CONTROL COMPONENTS

refrigerant flow from and to the compressor is the same. The reversing valve was used to change the direction of flow.

REVERSING VALVE OPERATION

The 4-way reversing valve is actually two valves that operate together. There is a main valve that actually controls the flow of refrigerant in the system, and a pilot valve that controls the operation of the main valve, figure 29-4. The force needed to operate the main valve is provided by the compressor. The valve shown in figure 29-4 has a sliding valve body that is used to control the flow of refrigerant through the system. In the illustration shown, the system is being used in the cooling cycle. Notice that on each side of the valve there is a small passage called an orifice. The orifice provides a path for a very small amount of refrigerant to flow. Notice also that there is a small capillary tube con-

nected from each end of the valve body to the pilot valve and a third capillary tube connected from the pilot valve to the suction line of the compressor. In the position shown, the plunger of the pilot valve is blocking the capillary from the left side of the main valve body. The capillary tube connected to the right side of the main valve is connected to the suction side of the compressor through the pilot valve. With the plunger of the pilot valve in this position, a high pressure is formed on the left side of the main valve and a low pressure is formed on the right side. The high pressure created on the left side of the main valve forces the main valve to slide to the right. With the main valve in this position, the discharge line of the compressor is connected to the outside coil and the inside coil is connected to the suction side of the compressor.

If the solenoid coil is energized, the plunger of the pilot valve will change to the position shown in figure 29-5. The plunger now blocks the capillary tube connected to the right side of the main

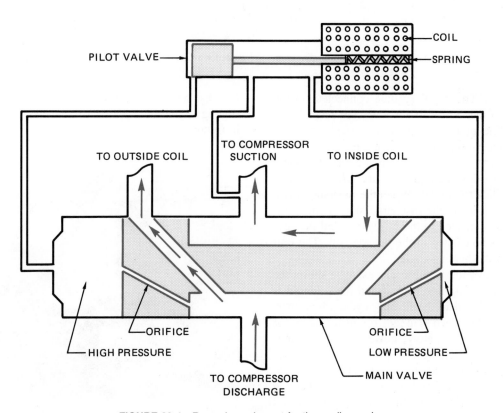

FIGURE 29-4 Reversing valve set for the cooling cycle

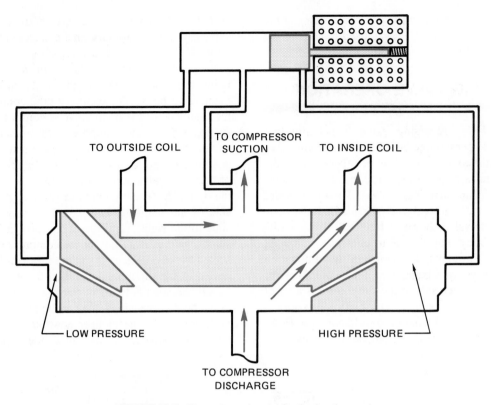

TO OUTSIDE COIL

TO COMPRESSOR SUCTION

TO INSIDE COIL

LOW PRESSURE

HIGH PRESSURE

TO COMPRESSOR DISCHARGE

FIGURE 29-5   Reversing valve set for the heating cycle

valve body. The capillary tube connected on the left side of the main valve is now connected to the suction line of the compressor through the pilot valve. This causes a high pressure to be created on the right side of the main valve and a low pressure on the left side. The high pressure forces the main valve to slide to the left. When the reversing valve is in this position, the discharge side of the compressor is connected to the inside coil and the suction line is connected to the outside coil. The unit is now in the heating cycle.

The valve illustrated in this example shows the valve is in the cooling cycle when the solenoid is de-energized and in the heating cycle when energized. This has been standard for many years for heat-pump systems. Now, however, some manufacturers are reversing this procedure. Some reversing valves are made in such a manner that when the valve is de-energized the unit is in the heating cycle. This was done so that valve failure would result in the unit being in the heating cycle. It is

felt that heat is necessary to life while air conditioning is not. A 4-way reversing valve is shown in figure 29-6.

FIGURE 29-6   Reversing valve

SECTION 4   CONTROL COMPONENTS

REVIEW QUESTIONS

1. What is a solenoid valve?

2. Why is it important not to reverse the connection of the inlet and outlet side of a solenoid valve?

3. What is used to cause the plunger to close when the solenoid coil is de-energized?

4. What is the function of a 4-way reversing valve?

5. What is the function of the pilot valve?

6. What is the function of the main valve?

7. What is actually used to change the position of the main valve from one setting to another?

UNIT 30
The Short-Cycle Timer

Short cycling is a condition that occurs when the compressor is restarted immediately after it has been turned off. This causes the compressor to restart against a high head pressure. Trying to restart a compressor in this manner can cause damage to the compressor, motor winding, or at the very least, open a circuit breaker or overload relay. After a compressor has been turned off, enough time should be permitted to pass to allow the pressure in the system to equalize before it is restarted.

CAUSES OF SHORT CYCLING

Short cycling can be caused by several situations. For example, a loose thermostat wire can cause a bad connection that will cause the compressor to alternately start and stop. A momentary interruption of the power line can cause the compressor to stop and then restart when power is restored. People can also cause short cycling. Assume, for example, that the air-conditioning system is in operation. Now assume that someone changes the thermostat setting and causes the thermostat contact to open and stop the compressor. Now assume that the person changes their mind and again changes the thermostat so that the compressor tries to restart. Regardless of the reason or causes of short cycling, it should be avoided whenever possible.

THE SHORT-CYCLING TIMER

The short-cycling timer is a cam operated, motor driven, on delay timer. A photograph of a short-cycling timer is shown in figure 30-1. This timer is used in conjunction with a relay generally referred to as a holding relay. The timer contains a set of double-pole double-throw contacts (DPDT). A basic schematic of a short-cycling timer circuit

FIGURE 30-1 Short-cycle timer

is shown in figure 30-2. Notice the two sets of double-throw contacts labeled A and B. The dashed line between the contacts indicates mechanical connection so that they operate together. Notice also the holding relay labeled HR. The circuit is controlled by the operation of the thermostat.

CIRCUIT OPERATION

To understand the operation of this circuit, refer to the schematic shown in figure 30-3. The arrows indicate the paths for current flow. Notice there is a path from line L1 through the primary of the control transformer and back to L2. This provides 24 volts for the operation of the control circuit. When the thermostat contacts close, a circuit is provided through the coil of the control relay (CR).

This causes contact CR to close and provides a current flow path to the short-cycle timer. Notice there are two paths of current flow at the timer. The current enters the A1 terminal and flows to the B1 contact terminal. The current can then flow through the contact to terminal B and then to the timer motor. There is also a current path from terminal A1 to A. The current can then flow from A to the holding relay (HR).

The holding relay energizes and changes both HR contacts as shown in figure 30-4. The now closed HR1 contact is used as a holding contact. Notice that current is also flowing through the timer motor. The timer motor is geared to permit a delay of about 3 minutes before the contacts change position.

Figure 30-5 illustrates the operation of the circuit when the timer contacts change position. No-

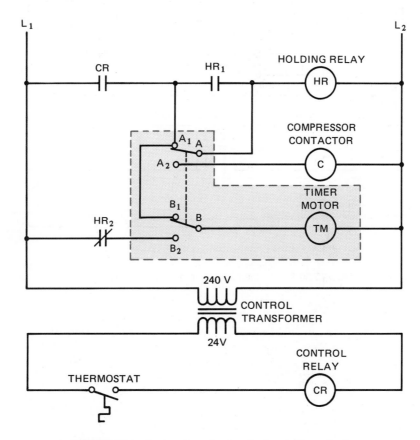

FIGURE 30-2 Basic schematic of a short-cycle timer circuit

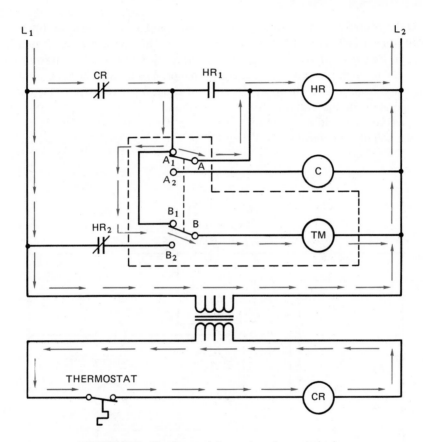

FIGURE 30-3 The thermostat energizes the control relay.

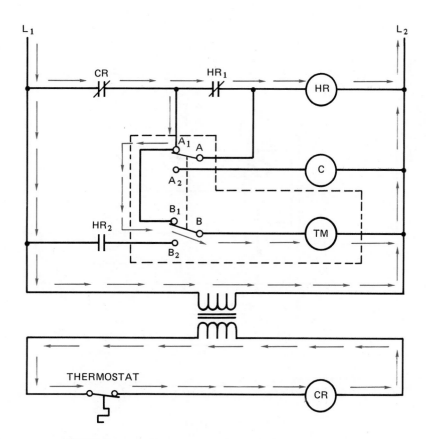

FIGURE 30-4 The holding relay and the timer motor are energized.

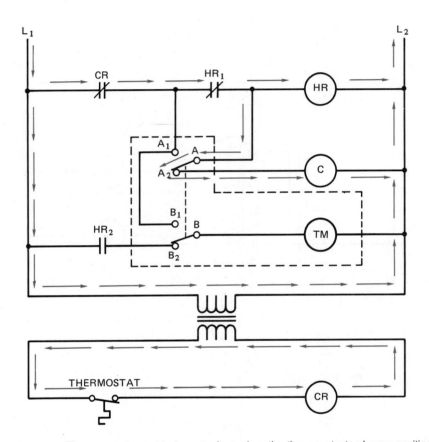

FIGURE 30-5 The compressor contactor energizes when the timer contacts change position.

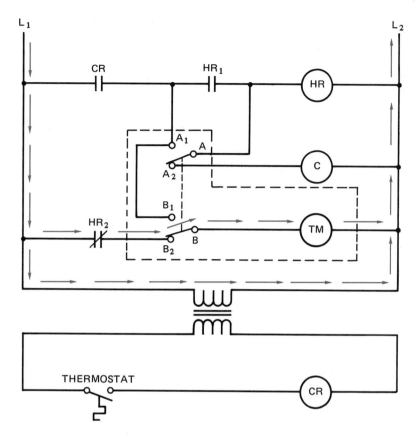

FIGURE 30-6 When the thermostat de-energizes, current is provided to the timer motor to reset the contacts.

tice that current now flows through the closed CR contact and the closed HR contact to contact A. Current can now flow to contact A2 and then to the compressor contactor. When the compressor contactor energizes, it connects the compressor to the line. Notice that contact B2 is connected to the now open HR2 contact. Since there is no current flow to the timer motor, the timing operation is stopped for as long as the thermostat maintains a circuit to CR relay.

After the thermostat has been satisfied, it will reopen and de-energize CR relay coil. This causes CR contact to open and de-energize HR relay coil. When HR relay de-energizes, HR2 contacts will again close and current flow is provided through the B2 and B contact to the timer motor, figure 30-6. The timer motor now operates and resets the contacts to their original position as shown in figure 30-2. The circuit is now ready for another operation sequence. If the thermostat momentarily opens and then recloses, or if there is a momentary loss of power, the holding relay will de-energize and the timer will have to time out before the compressor can be reconnected to the line.

REVIEW QUESTIONS

1. What is short cycling?

2. What is used to provide the timing operation for the short-cycle timer?

3. What type of contacts are used in the short-cycle timer?

4. How many and what type of contacts must the holding relay have?

5. What does the dashed line drawn between the two sets of timer contacts represent?

UNIT 31

Methods of Sensing Temperature

In the air-conditioning and refrigeration field, the ability to sense and measure temperature is of great importance. There are numerous methods used to sense the temperature. In fact, there has probably been more emphasis on the ability to measure temperature than any other quantity. This unit will deal with some of these methods.

EXPANSION OF METAL

A very common and reliable method for sensing temperature is by the expansion of metal. It has been known for many years that metal expands when heated. The amount of expansion is proportional to two things:

1. The type of metal used,
2. The amount of heat.

Consider the metal bar shown in figure 31-1. When the bar is heated, its length expands. When the metal is permitted to cool, it will contract. Although the amount of the movement due to contraction and expansion is small, a simple mechanical principle can be used to increase the amount of movement, figure 31-2.

The metal bar is mechanically held at one end. This permits the amount of expansion to be in only one direction. When the metal is heated and the bar expands, it pushes against the mechanical arm. A small movement of the bar causes a great amount

of movement in the mechanical arm. This increased movement of the arm can be used to indicate the temperature of the bar, or it can be used to operate a switch as shown. It should be understood that illustrations are used to convey a principle. In actual practice, the switch shown in figure 31-2 would be spring loaded to provide a ''snap'' action for the contacts. Electrical contacts must never be permitted to open or close slowly. This produces poor contact pressure and will cause the contacts to burn, or cause erratic operation of the equipment they are intended to control. A device that uses this principle is one type of starting relay known as the hot-wire relay. This starting relay was covered in an earlier chapter.

Another very common device that operates on the principle of expansion and contraction of metal is the mercury thermometer. Mercury is a metal that remains in a liquid state at room temperature. If the mercury is confined in a narrow glass tube as shown in figure 31-3, it will rise up the tube as it expands due to heat. If the tube is calibrated cor-

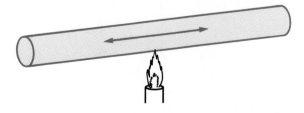

FIGURE 31-1 Metal expands when heated.

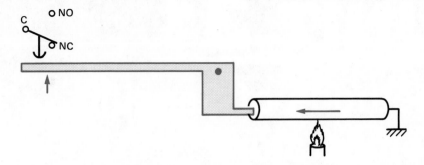

FIGURE 31-2 Expanding metal operates a set of contacts.

rectly, it provides an accurate measurement for temperature.

THE BIMETAL STRIP

The bimetal strip is another device that operates by the expansion of metal. It is probably the most common heat-sensing device used in the production of thermometers and thermostats. The bimetal strip is made by bonding two dissimilar types of metal together, figure 31-4. Since these metals are not alike, they have different expansion rates. This causes the strip to bend or wrap when heated, figure 31-5.

The bimetal strip is often formed into a spiral shape as shown in figure 31-6. The spiral permits a longer bimetal strip to be used in a small space. The longer the strip is, the more movement that will be produced by a change of temperature. If one end of the strip is mechanically held, and a pointer attached to the center of the spiral, a change in temperature will cause the pointer to rotate. If a calibrated scale is placed behind the pointer, it becomes a thermometer. If the center of the spiral is held, and a contact is attached to the end of the bimetal strip, it becomes a thermostat. As stated previously, electrical contacts cannot be permitted to open or close slowly. This type of thermostat uses a small permanent magnet to provide a snap

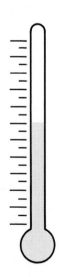

FIGURE 31-3 A mercury thermometer operates by the expansion of metal.

FIGURE 31-4 A bimetal strip

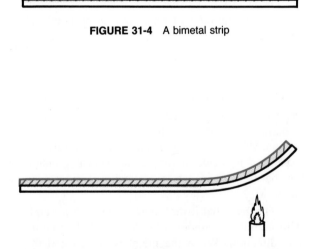

FIGURE 31-5 A bimetal strip warps with a change of temperature.

SECTION 4 CONTROL COMPONENTS

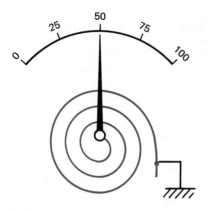

FIGURE 31-6 A bimetal strip used as a thermometer

action for the contact, figure 31-7. When the moving contact reaches a point that is close to the stationary contact, the magnet attracts the metal strip and causes a sudden closing of the contacts. When the bimetal strip cools, it attempts to pull itself away from the magnet. When the force of the bimetal strip becomes strong enough, it overcomes the force of the magnet and the contacts snap open. This type of thermostat is inexpensive and has been used in homes for many years.

THE THERMOCOUPLE

The thermocouple is made by joining two dissimilar metals together at one end. When the joined end of the thermocouple is heated, a voltage is produced at the opposite end, figure 31-8. The amount of voltage produced is proportional to:

1. The types of metals used to produce the thermocouple,

2. The difference in temperature of the two junctions.

The chart in figure 31-9 shows common types of thermocouples. The metals the thermocouples are constructed from are shown as well as their normal temperature range.

The amount of voltage produced by a thermocouple is small, generally on the order of millivolts (1 millivolt = .001 volt). The polarity of the voltage of a thermocouple is determined by the temperature. For example, a type "J" thermocouple produces zero volts at about 32 degrees Fahrenheit. At temperatures above 32 degrees, the iron wire is positive and the constantan wire is negative. At a temperature of 300 degrees, this thermocouple will produce a voltage of about 7.9 millivolts. At a temperature of −300 degrees, it produces a voltage of about −7.5 millivolts. This indicates that at temperatures below 32 degrees Fahrenheit the iron wire becomes the negative lead and the constantan wire becomes the positive-voltage lead.

Since thermocouples produce such low voltages, they are often connected in series as shown in figure 31-10. This series connection permits the voltages to add and produce a higher output voltage. This connection is known as a thermopile.

RESISTANCE TEMPERATURE DETECTORS

The resistance temperature detector (RTD) is made of platinum wire. The resistance of platinum changes greatly with temperature. When platinum

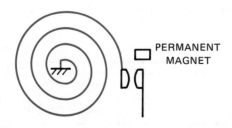

FIGURE 31-7 A bimetal strip used to operate a set of contacts

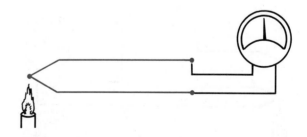

FIGURE 31-8 A thermocouple produces a voltage when the two ends are at different temperatures.

TYPE	MATERIAL		DEGREES F	DEGREES C
J	Iron	Constantan	−328 to +32 +32 to +1432	−200 to 0 0 to 778
K	Chromel	Alumel	−328 to +32 +32 to +2472	−200 to 0 0 to 1356
T	Copper	Constantan	−328 to +32 +32 to +752	−200 to 0 0 to 400
E	Chromal	Constantan	−328 to +32 +32 to 1832	−200 to 0 0 to 1000
R	Platinum 13% Rhodium	Platinum	+32 to +3232	0 to 1778
S	Platinum 10% Rhodium	Platinum	+32 to +3232	0 to 1778
B	Platinum 30% Rhodium	Platinum 6% Rhodium	+992 to +3352	533 to 1800

FIGURE 31-9   Thermocouple chart

is heated, its resistance increases at a very predictable rate. This makes the RTD an ideal device for measuring temperature very accurately. RTDs are used to measure temperatures that range from −328 to +1166 degrees Fahrenheit (−200 to +630C). RTDs are made in different styles to perform different functions. Figure 31-11 illustrates a typical RTD used as a probe. A very small coil of platinum wire is encased inside a copper tip. Copper is used to provide good thermal contact. This permits the probe to be very fast-acting. The chart in figure 31-12 shows resistance versus temperature for a typical RTD probe. The temperature is given in degrees Celsius and resistance is given in ohms.

THERMISTORS

The term *thermistor* is derived from the words thermal resistor. Thermistors are actually thermally-sensitive semi-conductor devices. There are two basic types of thermistors. One type has a neg-

FIGURE 31-10   Thermocouple

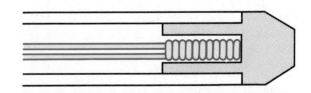

FIGURE 31-11   Resistance temperature detector

SECTION 4   CONTROL COMPONENTS

DEGREES C	RESISTANCE
0	100
50	119.39
100	138.5
150	157.32
200	175.84
250	194.08
300	212.03
350	229.69
400	247.06
450	264.16
500	280.93
550	297.44
600	313.65

FIGURE 31-12 Temperature and resistance for a typical RTD

ative temperature coefficient (NTC) and the other has a positive temperature coefficient (PTC). A thermistor that has a negative temperature coefficient will decrease its resistance as the temperature increases. A thermistor that has a positive temperature coefficient will increase its resistance as temperature increases. The NTC thermistor is the most widely used.

Thermistors are highly nonlinear devices. For this reason they are difficult to use for measuring temperature. Devices that measure temperature with a thermistor must be calibrated for the particular type of thermistor being used. If the thermistor is ever replaced, it has to be an exact replacement or the circuit will no longer operate correctly. Because of their nonlinear characteristic, thermistors are often used as set point detectors as opposed to actual temperature measurement. A set point detector is a device that activates some process or circuit when the temperature reaches a certain level. For example, assume a thermistor has been placed inside the stator of a motor used to operate a compressor. If the motor should become overheated, the windings of the motor could be severely damaged or destroyed. The thermistor can be used to detect the temperature of the windings. When the resistance of the thermistor falls to a certain level, NTC type, a set of contacts connected in series with the motor starter coil of the compressor, opens. When the compressor motor starter de-energizes,

the compressor is disconnected from the power line. Thermistors can be operated in temperatures that range from about −100 to +300 degrees Fahrenheit.

THE PN JUNCTION

Another device that has the ability to measure temperature is the PN junction or diode. The diode is becoming a very popular device for measuring temperature because it is accurate and linear.

When a silicon diode is used as a temperature sensor, a constant current is passed through the diode. Figure 31-13 shows this type of circuit. In this circuit, resistor R1 limits the current flow through the transistor and the sensor diode. The value of R1 also determines the amount of current that will flow through the diode. Diode D1 is a 5.1-volt zener diode used to produce a constant voltage between the base and emitter of the PNP transistor. Resistor R2 limits the amount of current flow through the zener diode and the base of the transistor. Diode D2 is a common silicon diode. It is being used as the temperature sensor for the circuit. If a digital voltmeter is connected across the diode, a voltage drop between .8 and 0 volts can be seen. The amount of the voltage drop is determined by the temperature of the diode.

If the diode is subjected to a lower temperature, say by touching it with a piece of ice, it will

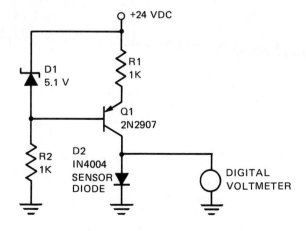

FIGURE 31-13 Constant current generator

be seen that the voltage drop of the diode will increase. If the temperature of the diode is increased by holding it between two fingers or bringing a hot soldering iron near it, its voltage drop will decrease. Notice that the diode has a negative temperature coefficient. As its temperature increases, its voltage drop becomes less. The circuit shown in figure 31-14 can be used as a set point detector. The operation of the circuit is as follows:

A bridge rectifier and a center-tapped transformer are used to produce an above- and below-ground power supply. If ground is considered as zero volts, the positive output of the bridge will be positive with respect to ground and the negative output of the bridge will be negative with respect to ground. Capacitors C1 and C2 are used to filter the DC output voltage of the rectifier. Notice that capacitor C1 has its positive lead connected to ground and C2 has its negative lead connected to ground. The positive output of the rectifier will

produce a voltage that is about +9 volts compared to ground, and the negative output will produce a voltage that is about −9 volts compared to ground.

Diode D1 is a light-emitting diode connected in the forward direction. In this circuit, the LED is used as a low-voltage zener diode. Since the LED has a constant voltage drop of about 1.7 volts, it can be used to provide a constant voltage. Resistor R1 limits the current flow through the diode and the sensor resistor. Resistor R2 limits current flow through the LED and the base of the transistor. Notice this is the same constant current generator circuit shown in figure 31-13 with the exception of the LED being used as the zener diode.

Transistor Q2, resistors R5 and R4, and LED D4 form another constant current generator circuit. Notice this generator is connected to an LED, D5. In this circuit, D5 is used to provide a low-voltage reference source for the operational amplifier. When a light-emitting diode is connected to a constant

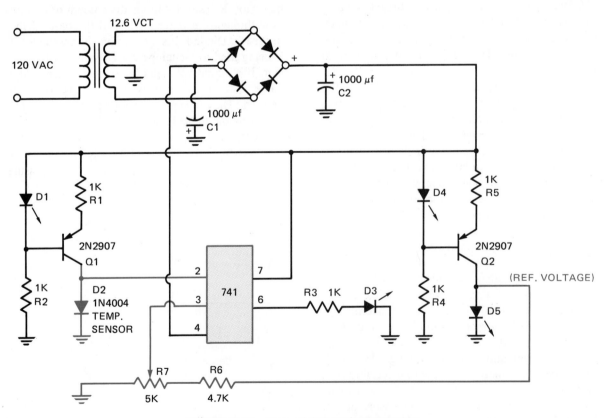

FIGURE 31-14 Set level detector for temperature

current source, its voltage drop is very stable. This makes it an ideal choice when a steady reference voltage is needed. Resistors R6 and R7 are used to form a voltage divider. Resistor R5 is a 5000-ohm variable resistor that has a voltage drop across its entire resistance of about 1 volt. The wiper tap of this resistor is connected to the noninverting input of the 741. Since resistor R5 has a voltage drop of only 1 volt across its resistance, the full range of the wiper will adjust the voltage applied to the noninverting input between 1 volt and 0. This is done to make adjustment of the detector circuit easy. Since the voltage drop of diode D2 will never be greater than .8 volts, resistor R7 can adjust the entire range over which the detector can operate.

Diode D3 is used as an output indicator. When the output is low, D3 will be turned off. When the output of the op amp goes high, D3 will be turned on. Diode D3 is used only as an indicator in this circuit. The output of the op amp could be used to operate the input of a transistor or a solid-state relay. The transistor or relay could be used to operate almost anything desired. Resistor R3 limits the current flow through D3.

To understand the operation of this circuit, assume that resistor R7 has been adjusted to a point that the output of the op amp is off or low. This means that the voltage applied to the inverting input, pin 2, is more positive than the voltage set at pin 3. If the temperature of diode D2 is increased, its voltage drop will decrease. When the temperature of the sensor diode becomes high enough, its voltage drop will be less than the voltage set at the noninverting input. When the voltage applied to pin 3 becomes more positive than the voltage applied to pin 2, the output of the op amp will go high or

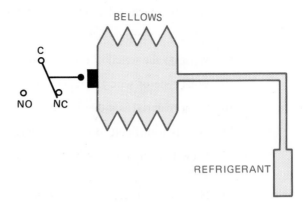

FIGURE 31-15 Bellows contracts and expands with a change of refrigerant pressure.

turn on. Adjustment of resistor R7 permits the detector to be used over a wide range of temperatures.

EXPANSION DUE TO PRESSURE

Another common method of measuring temperature is by the increase of pressure of some chemicals. Refrigerant, for example, increases pressure as temperature increases. If a simple bellows is connected to a line containing refrigerant, figure 31-15, the bellows will expand as the pressure inside the sealed system increases. When the surrounding temperature decreases, the pressure inside the system decreases, and the bellows contracts. When the bellows is made to operate a set of contacts, it is generally referred to as a bellows-type thermostat.

REVIEW QUESTIONS

1. Should a metal bar be heated or cooled to make it expand?

2. What type of metal remains in a liquid state at room temperature?

3. How is a bimetal metal strip made?

4. Why are bimetal strips often formed into a spiral shape?

5. Why should electrical contacts never be permitted to open or close slowly?

6. What two factors determine the amount of voltage produced by a thermocouple?

7. What is a thermopile?

8. What do the letters RTD stand for?

9. What type of wire are RTDs made of?

10. What material is a thermistor made of?

11. Why is it difficult to measure temperature with a thermistor?

12. If the temperature of a NTC thermistor increases, will its resistance increase or decrease?

13. How can a silicon diode be made to measure temperature?

14. Assume that a silicon diode is being used as a temperature detector. If its temperature increases, will its voltage drop increase or decrease?

15. What is an above- and below-ground power supply?

SECTION 5

Control Circuits

UNIT 32
Schematics and Wiring Diagrams

Schematics and wiring diagrams are the written language of control circuits. It will be impossible for a service technician to become proficient in troubleshooting electrical faults if he or she cannot read and interpret electrical diagrams. Learning to read electrical diagrams is not as difficult as many people first believe it to be. Once a few basic principles are understood, reading schematics and wiring diagrams will become no more difficult than reading a newspaper.

TWO-WIRE CIRCUITS

Control circuits are divided into two basic types, the two-wire and the three-wire. Figure 32-1 shows a simple two-wire control circuit. In this circuit, a simple switch is used to control the power applied to a small motor. If the switch is open, there is no complete path for current flow, and the motor will not operate. If the switch is closed, power is supplied to the motor, and it then operates.

THREE-WIRE CIRCUITS

Three-wire control circuits are used because they are more flexible than two-wire circuits. Three-wire circuits are characterized by the fact that they are operated by a magnetic relay or motor starter.

These circuits are generally controlled by one or more pilot devices.

ELECTRICAL SYMBOLS

When a person first learns to read, he or she learns a set of symbols that are used to represent different sounds. This set of symbols is generally referred to as the alphabet. When learning to read electrical diagrams, it is necessary to learn the symbols used to represent different devices and components. The symbols shown below are commonly used on control schematics and wiring diagrams. These are not all the symbols used. Unfortunately, there is no set standard for the use of electrical symbols. Most of these symbols are approved by the National Electrical Manufacturers Association (NEMA). These symbols are as follows:

1. Normally closed push button. Generally used to represent a stop button.

2. Normally open push button. Generally used to represent a start button.

3. Double-acting push button. Contains both normally closed and normally open contacts on one push button.

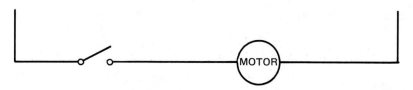

FIGURE 32-1 Two-wire control circuit

4. Double-acting push button drawn differently but meaning the same as number 3.

5. Double-acting push button. The dashed line indicates mechanical intertie. This means that when one button is pushed, the other moves at the same time.

6. Single-pole single-throw switch (SPST).

7. Single-pole double-throw switch (SPDT). Notice this switch has only one pole, the switch arm, but it has two stationary contacts. In the diagram, the switch arm makes contact with the upper stationary contact. When the switch is thrown, contact will be made between the switch arm and the lower stationary contact. The switch arm or pole of the switch is generally referred to as the common because it can make contact to either of the two stationary contacts.

8. Double-pole single-throw switch (DPST). Notice the dashed line, which indicates mechanical intertie between the two switch arms.

9. Double-pole double-throw switch (DPDT).

10. Off-Automatic-Manual control switch. This switch is basically a single-pole double-throw that has a center off position.

11. Normally open relay contact.

12. Normally closed relay contact.

13. Fuse

14. Fuse

15. Transformer

16. Coil

17. Coil (Generally used to represent the coil of a relay or motor starter in a control schematic.)

18. Pilot light or lamp

19. Lamp or light bulb

20. Thermal heater element

21. Thermal heater element. (Generally used to represent the overload heater element in a motor control circuit.)

22. Solenoid coil

23. Fixed resistor

24. Variable resistor

25. Variable resistor

26. Single-pole circuit breaker

27. Double-pole circuit breaker

 28. Capacitor

 29. Normally closed float switch

 30. Normally open float switch

 31. Normally closed pressure switch

 32. Normally open pressure switch

 33. Normally closed temperature switch. (Normally closed thermostat)

 34. Normally open temperature switch.

 35. Normally closed flow switch. This symbol is used to represent both liquid- and air-sensing flow switches.

 36. Normally open flow switch.

 37. Normally closed limit switch.

 38. Normally open limit switch

 39. Normally closed ON-DELAY timer contact. Often shown on schematics as DOE, which stands for Delay On Energize.

 40. Normally open ON-DELAY timer contact.

 41. Normally closed OFF-DELAY timer contact. Often shown on schematics as DODE, which stands for Delay on De–Energize.

 42. Normally open OFF-DELAY timer contact

 43. Battery

 44. Electrical ground

 45. Mechanical ground

 46. Wires crossing without connection

 47. Wires crossing without connection

 48. Wires connecting. The dot in the center of the cross is known as a *node*. This is used to indicate connection.

49. Rotary switch

The contact symbols shown are standard and relatively simple to understand. There can be instances, however, in which symbols can be used to show something that is not apparent. For example, the symbol for a normally closed pressure switch is shown in figure 32-2. Notice that this symbol not only shows the movable arm making contact with the stationary contact, but it also shows the movable arm drawn above the stationary contact. In figure 32-3, the contact arm is shown not

FIGURE 32-2 Normally closed pressure switch

FIGURE 32-3 Normally closed held-open pressure switch

FIGURE 32-4 Normally open pressure switch

FIGURE 32-5 Normally open held-closed pressure switch

making connection with the stationary contact. This symbol, however, is not a normally open contact symbol because the contact arm is drawn above the stationary contact. This symbol indicates a normally closed held-open pressure switch. This symbol is indicating that the switch is actually connected as a normally closed switch, but pressure is used to keep the contact open. If pressure decreases to a certain point, the switch contact will close.

Figure 32-4 shows a normally open pressure switch. Notice that the contact arm is drawn below the stationary contact. Figure 32-5 shows the same symbol except that the movable arm is making connection with the stationary contact. This symbol represents a normally open held-closed pressure switch. This switch symbol indicates that the pressure switch is wired normally open, but pressure switch is wired normally open, but pressure holds the contact closed. If the pressure decreases to a certain level, the switch will open and break connection to the rest of the circuit.

SCHEMATIC DIAGRAMS

Schematic diagrams show components in their electrical sequence without regard for physical location. Schematic diagrams are used to trouble-shoot and install control circuits. Schematics are generally easier to read and understand than wiring diagrams.

WIRING DIAGRAMS

Wiring diagrams show components mounted in their general location with connecting wires. A wiring diagram is used to represent how the circuit generally appears. To help illustrate the differences between wiring diagrams and schematics, a basic control circuit will first be explained as a schematic and then shown as a wiring diagram.

READING SCHEMATIC DIAGRAMS

To read a schematic diagram, a few rules must first be learned. Commit the following rules to memory:

1. Reading a schematic diagram is similar to reading a book. It is read from left to right and from top to bottom.
2. Electrical symbols are always shown in their off or de-energized position.
3. Relay contact symbols are shown with the same numbers or letters that are used to designate the

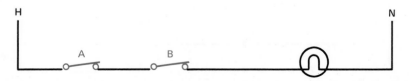

FIGURE 32-6 Components used to perform the function of stop are normally closed and connected in series.

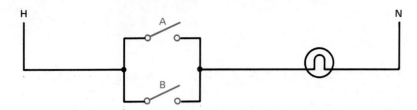

FIGURE 32-7 Components used to perform the function of start are normally open and connected in parallel.

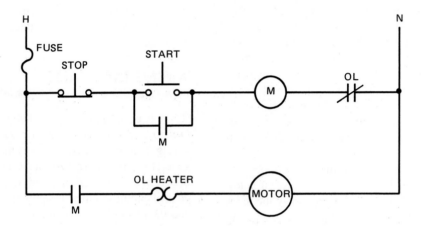

FIGURE 32-8 Start-stop push-button control circuit

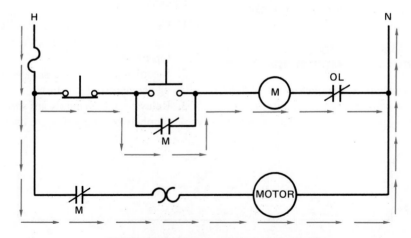

FIGURE 32-9 Current path through the circuit

SECTION 5 CONTROL CIRCUITS

relay coil. All contact symbols that have the same number or letter as a coil are controlled by that coil regardless of where in the circuit they are located.

4. When a relay is energized, or turned on, all of its contacts change position. If a contact is shown as normally open, it will close when the coil is energized. If the contact is shown normally closed, it will open when the coil is turned on.

5. There must be a complete circuit before current can flow through a component.

6. Components used to provide a function of stop are generally wired normally closed and connected in series. Figure 32-6 illustrates this concept. Both switches A and B are normally closed and connected in series. If either switch is opened, connection to the lamp will be broken and current will stop flowing in the circuit.

7. Components used to provide the function of start are generally wired normally open and connected in parallel. In figure 32-7, switches A and B are normally open and connected in parallel with each other. If either switch is closed, a current path will be provided for the lamp and it will turn on.

The circuit to be discussed is a basic control circuit used throughout industry. Figure 32-8 shows a start-stop push button circuit. This schematic shows both the control circuit and the motor circuit. Schematic diagrams do not always show both control and motor connections. Many schematic diagrams show only the control circuit.

Notice in this schematic that there is no complete circuit to M motor starter coil because of the open start push button and open M auxiliary contacts. There is also no connection to the motor because of the open-M load contacts. The open-M contacts connected in parallel with the start button are small contacts intended to be used as part of the control circuit. This set of contacts is generally referred to as the *holding*, *sealing*, or *maintaining* contacts. These contacts are used to provide a continued circuit to the M coil when the start button is released.

The second set of M contacts is connected in series with the overload heater element and the motor, and are known as *load* contacts. These contacts are large and designed to carry the current needed to operate the load. Notice that these contacts are normally open and there is no current path to the motor.

When the start button is pushed, a path for current flow is provided to the M-motor starter coil. When the M coil energizes, both M contacts close, figure 32-9. The small auxiliary contact provides a continued current path to the motor starter coil when the start button is released and returns to its open position. The large M load contact closes and provides a complete circuit to the motor and the motor begins to run. The motor will continue to operate in this manner as long as the M coil remains energized.

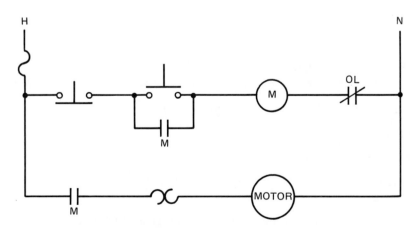

FIGURE 32-10 The stop button breaks the circuit.

If the stop button is pushed, figure 32-10, the current path to the M coil is broken and the coil de-energizes. This causes both M contacts to return to their normally open position. When M holding contacts open, there is no longer a complete circuit provided to the coil when the stop button is returned to its normal position. The circuit remains in the off position until the start button is again pushed.

Notice that the overload contact is connected in series with the motor starter coil. If the overload contact should open, it has the same effect as pressing the stop button. The fuse is connected in series with both the control circuit and the motor. If the fuse should open, it has the effect of disconnecting power from the line.

A wiring diagram for the start-stop push-button circuit is shown in figure 32-11. Although this diagram looks completely different, it is electrically the same as the schematic diagram. Notice the push button symbols indicate double-acting push buttons. The stop button, however, uses only the normally closed section and the start button uses only the normally open section. The motor starter shows three load contacts and two auxiliary contacts. One auxiliary contact is open and one is closed. Notice that only the open contact has been used.

The overload unit shows two different sections. One section contains the thermal heater element connected in series with the motor, and the normally closed contact is connected in series with the coil of the M-motor starter.

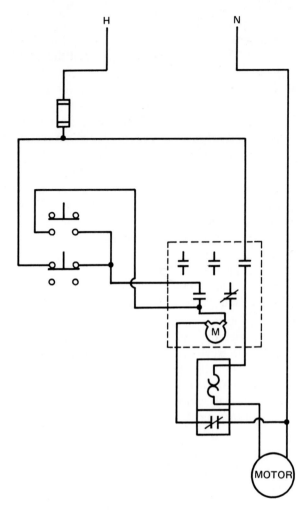

FIGURE 32-11 Wiring diagram of start-stop push-button control circuit

REVIEW QUESTIONS

1. What are the two basic types of motor controls?

2. Define a schematic diagram.

3. Define a wiring diagram.

4. Components used for the function of stop are generally wired _____ _____ and connected in _____.

5. Components used for the function of start are generally wired _____ _____ and connected in _____.

SECTION 5   CONTROL CIRCUITS

6. When reading a schematic diagram, are the components shown in their energized or de-energized position?

7. What does this symbol represent?

8. What does this symbol represent?

9. What does a dashed line drawn between components represent?

10. What is an auxiliary contact?

UNIT 33

Developing Wiring Diagrams

In this unit, two schematic diagrams are discussed, including their operation and development into wiring diagrams. Developing a wiring diagram from a schematic is the same basic procedure that is followed when installing a control system. Understanding this process is also a great advantage when troubleshooting existing circuits.

DEVELOPING CIRCUIT #1

The first circuit discussed is shown in figure 33-1. In this circuit, a fan motor is controlled by relay FR (Fan Relay). The circuit is so designed that a switch can be used to turn the circuit completely off, operate the fan manually, or permit the

fan to be operated by a thermostat. If the control switch is moved to the "MAN" position as shown in figure 33-2, a complete circuit is provided to the fan relay coil. Then the relay energizes, FR contact closes and connects the motor to the line. This setting permits the fan to be operated at any time, regardless of the condition of the thermostat.

If the control switch is moved to the "AUTO" position as shown in figure 33-3, the fan will be controlled by the action of the thermostat. When the temperature increases to a predetermined level, the thermostat contact will close. This completes a circuit to FR coil. When coil FR energizes, the FR contact closes and connects the fan motor to the line. When the temperature decreases sufficiently, the thermostat contact opens and breaks the circuit

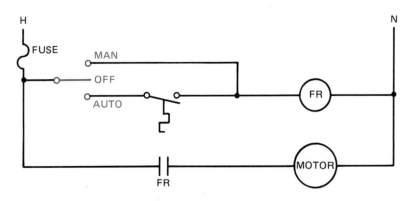

FIGURE 33-1 Fan control circuit

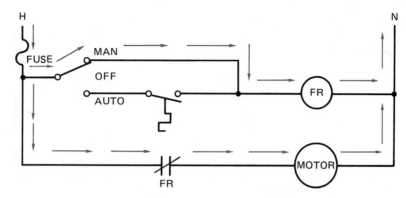

FIGURE 33-2 Fan relay coil is energized by control switch.

to FR coil. When FR coil de-energizes, FR contact opens and disconnects the motor from the line.

This schematic will now be developed into a wiring diagram. To aid in the connection of this circuit, a simple numbering system will be used. To use this numbering system, the following rules will be followed:

1. All components connected to the same line will receive the same number.
2. Any time a component is gone through, the number will change.
3. A set of numbers can be used only once.

Figure 33-4 shows the numbers places on the schematic. Notice that a 1 is placed at the incoming power line and a 1 is also placed at one side of the fuse. Since the fuse is a component, the number must change on the other side of it. Therefore, the fuse has a 2 on the other side. There is

also a 2 placed beside the common terminal of the OFF-MANUAL-AUTOMATIC switch, and a 2 placed beside one side of FR contact. Notice that all of these components have the same number because there is no break between them.

The AUTO side of the switch has been numbered 3, and one side of the thermostat has also been numbered 3. The other side of the thermostat is numbered 4, the MAN side of the switch is numbered 4, and one side of FR coil is numbered 4. The other side of the coil has been numbered 5, the neutral line is numbered 5, and one side of the motor is numbered 5. The other side of the motor is numbered 6, and the other side of FR contact is numbered 6.

Notice that all the points that are electrically connected together have the same number. Notice also that no set of numbers was used more than once.

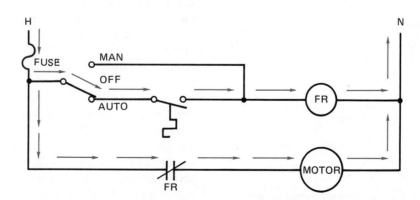

FIGURE 33-3 Fan relay is controlled by the thermostat.

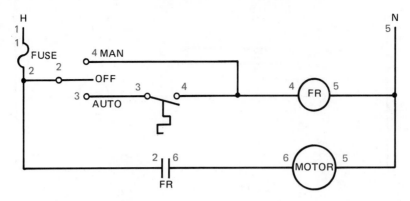

FIGURE 33-4 Schematic is numbered to aid in connection of the circuit.

The components of the system are shown in figure 33-5. Notice that numbers have been placed beside certain components. These numbers correspond to the numbers in the schematic. For example, the fuse in the schematic is shown with a 1 on one side and a 2 on the other side. The fuse

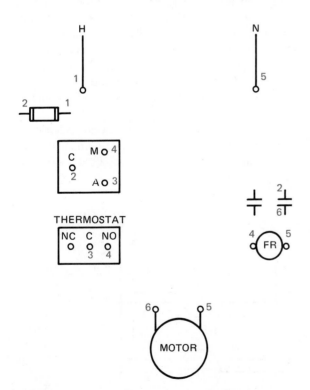

FIGURE 33-5 Circuit components are numbered with the same numbers that appear on the schematic.

in the wiring diagram is shown with a 1 on one side and a 2 on the other side. Notice the OFF-MANUAL-AUTOMATIC switch shown on the schematic. The common terminal is numbered 2, the MAN terminal is numbered 4 and the AUTO terminal is numbered 3. Now notice the same switch on the wiring diagram. The common terminal is numbered 2, the MAN terminal is numbered 4, and the AUTO terminal is numbered 3. The thermostat in the schematic has been numbered 3 on one terminal and 4 on the other terminal. The thermostat shown in the wiring diagram has three terminals. One terminal is common, one terminal is marked NC and the other terminal is marked NO. This is a common arrangement for many control components. This shows that the thermostat is a single-pole double-throw switch. Since the thermostat shown in the schematic is normally open, the 3 will be placed beside the common terminal, and the 4 will be placed beside the NO contact. Notice that one of the contacts on FR relay is numbered 2 on one side and 6 on the other side. FR coil is numbered 4 on one side and 5 on the other. One motor terminal is numbered 5 and the other is numbered 6.

Now that the component parts have been numbered with the same numbers as those used on the schematic, connection can be made easily and quickly. To connect the circuit, connect all the like numbers. For example, all the number 1s will connect together, all the number 2s will connect together, and so forth. The connected circuit is shown in figure 33-6.

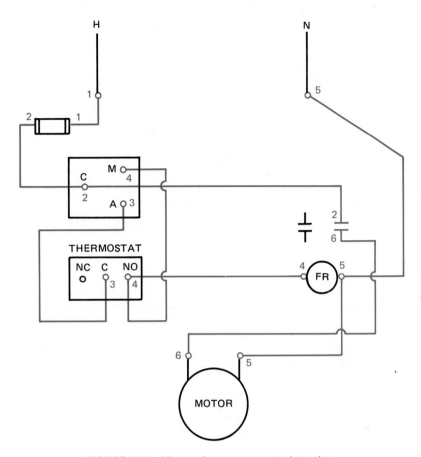

FIGURE 33-6 Like numbers are connected together.

DEVELOPMENT OF CIRCUIT #2

The schematic diagram for circuit #2 is shown in figure 33-7. Notice this schematic shows both the control circuit and the motor connection. This circuit is designed to turn off a compressor if the pressure in the system reaches a predetermined level. If the pressure becomes high enough to cause the pressure switch contacts to close, the compressor motor will not only be disconnected from the power line, but a warning light will also be turned on. Once the warning light has been turned on, the system must be manually reset by the service technician before the compressor can be restarted by the thermostat. The operation of the circuit is as follows:

When the thermostat contact closes, a circuit is completed to M motor starter coil. When M coil energizes, M contacts close and connect the compressor to the three-phase power line. When the temperature decreases, the thermostat contacts open and de-energize M coil. When M coil de-energizes, M contacts open and disconnect the compressor from the power line. Notice that in the normal action of this circuit, the compressor is controlled by the thermostat.

Now assume that the thermostat contacts are closed and that the compressor is connected to the power line. Also assume that the pressure in the system becomes too great and that the contacts of the pressure switch close. When the pressure switch contacts close, PSCR (Pressure Switch Control Relay) coil energizes. This causes both PSCR con-

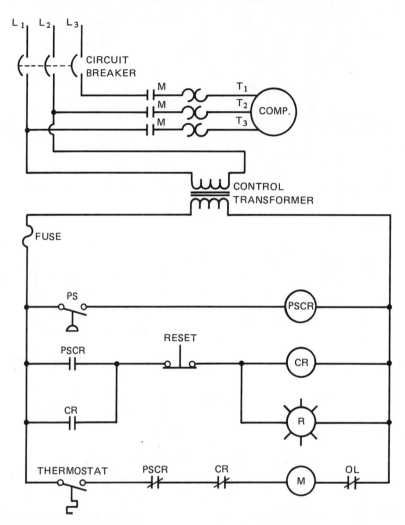

FIGURE 33-7 High pressure locks compressor off.

tacts to change position. When the normally closed PSCR contact opens, the circuit to M coil is broken. This causes the compressor to be disconnected from the line. When the normally open PSCR contact closes, a circuit is completed to CR (Control Relay) coil. When CR coil energizes, both CR contacts change position. The normally open CR contact closes to maintain a circuit to CR coil in the event that the pressure in the system decreases and opens the pressure switch contacts. This would cause PSCR relay to de-energize and return both PSCR contacts to their normal position. The normally closed CR contact will open. This prevents M coil from being energized by the thermostat if

PSCR contact should reclose. Notice that the warning light is connected in parallel with the coil of CR relay. The warning light will be turned on as long as CR relay coil is turned on. As long as CR relay is energized, the compressor cannot be restarted by the thermostat.

Now assume that the pressure in the system has returned to normal and the problem that caused the excessive pressure has been corrected. When the pressure switch contact reopened, PSCR coil de-energized and reset both PSCR contacts to their normal position. When the service technician presses the reset button, CR coil de-energizes and both CR contacts return to their normal positions. The cir-

SECTION 5 CONTROL CIRCUITS

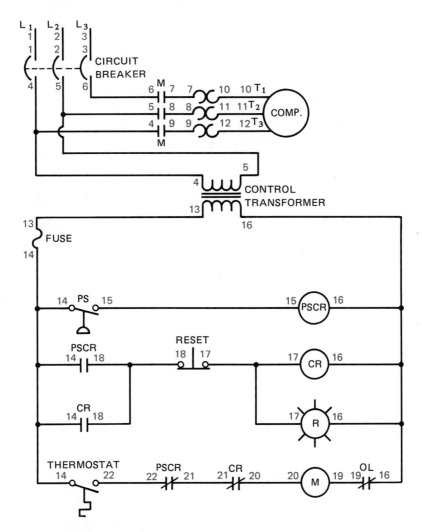

FIGURE 33-8 Schematic is numbered to aid in circuit connection.

cuit is now back in its original position and ready for normal operation.

This schematic diagram is now being developed into a wiring diagram. As before, the schematic is numbered in the same manner as the first example. The numbered schematic is shown in figure 33-8. Notice that all components that are electrically tied together have the same number. Also notice that no number set has been used more than once.

The control components are shown in figure 33-9. Notice that the numbers on the components correspond with like numbers on the schematic diagram. For example, on the schematic diagram

the primary of the control transformer is numbered 4 and 5. The secondary leads are numbered 13 and 16. Notice the same is true on the wiring diagram. Note the number of each component on the schematic and then find the corresponding number beside the proper component used on the wiring diagram.

Once the components of the wiring diagram have been numbered with the same numbers as those on the schematic, the circuit can be connected by connecting like numbers. The circuit connection is shown in figure 33-10. Again, notice that the wiring diagram appears to be completely different from the schematic, but both are the same electrically.

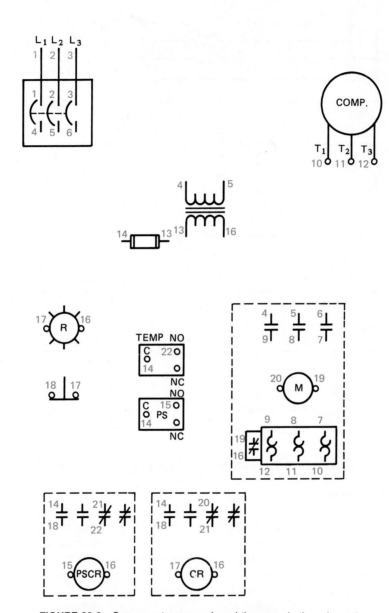

FIGURE 33-9 Components are numbered the same in the schematic.

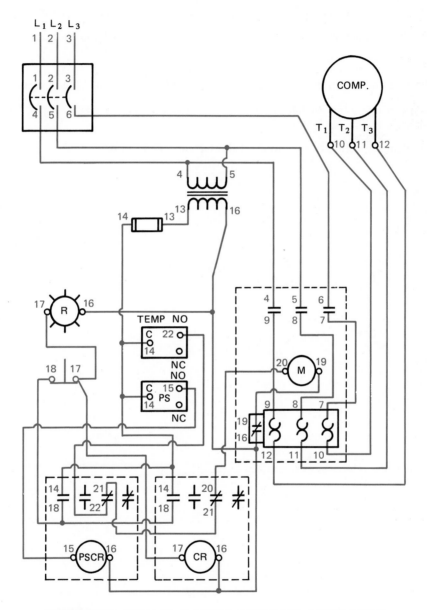

FIGURE 33-10 Connection is made by connecting like numbers together.

REVIEW QUESTIONS

Refer to circuit 33-1 for the following questions.

1. Explain the action of the circuit if the thermostat should fail to operate.

2. Explain the action of the circuit if FR contacts should become shorted together.

Refer to circuit 33-7 for the following questions.

3. Explain the action of the circuit if the overload (OL) contact should open.

4. Explain the action of the circuit if the pressure switch contacts should become shorted.

5. Explain the action of the circuit if the CR coil should open.

UNIT 34
Gas Burner Controls

The primary function of a gas burner control is to ensure that gas is not permitted to enter the system if it cannot be ignited in a safe manner. An accumulation of gas is extremely explosive and must be avoided. Several methods of igniting the main burner can be employed. The two most common in use today are the pilot light and high-voltage spark ignition.

PILOT LIGHT

Probably the oldest method of automatically igniting the main burner is with a pilot light. A pilot light is a small gas flame that burns continuously near the main burner. When gas is permitted to flow to the main burner, the pilot light ignites the fuel. If the pilot light should not be in operation when the gas is permitted to flow to the main burner, an accumulation of gas could result in an explosion. For this reason, the control system must have some means of sensing the presence of the pilot flame. If the pilot flame is not present, the main gas valve goes into safety shutdown and does not permit gas to be supplied to the main burner.

HIGH-VOLTAGE SPARK IGNITION

Many of the newer gas-operated appliances and heating systems use an electric arc to ignite the gas flame. This system uses less energy because it does not depend on a gas flame being present at all times. The electric arc is used only during the actual ignition sequence. When electric arc ignition is used, the gas-control system must be different also. Instead of sensing the presence of a pilot flame, the control system turns on the electric ignitor and permits gas to flow. If a flame is not detected in a short period of time, the control system turns off the flow of gas.

FLAME SENSORS

There are several methods used to sense the presence of a gas flame. One of the most common is with the use of a thermocouple. The thermocouple is a device that produces a voltage when heated. If the thermocouple is inserted in the gas flame as shown in figure 34-1, a voltage will be produced. The voltage produced by the thermocouple is used

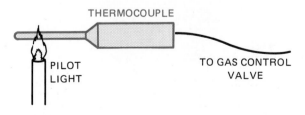

FIGURE 34-1 Thermocouple senses pilot flame.

to create a current flow through the coil of a solenoid. The current produces a magnetic field that holds the valve open. As long as the solenoid receives enough current, a valve is held open and gas is permitted to flow to the main burner. If the pilot light should go out, no voltage will be produced and the pilot valve will stop the flow of gas. It should be noted that the thermocouple has the ability to produce enough current to hold the valve open, but it cannot produce enough current to reopen the valve if it is closed. The pilot valve must be opened manually by pushing the pilot button located on the main valve. It should also be noted that some controls of this type are actually thermopiles and not thermocouples. Recall that a thermopile is a series connection of several thermocouples used to produce a higher voltage. When replacing a thermocouple, care must be taken to use the proper type. A thermocouple is shown in figure 34-2.

Another type of flame sensor uses pressure. This control is similar to the pressure type of thermostat. A refrigerant-filled bulb is located in the pilot flame. When the refrigerant is heated, a pressure is produced that holds the pilot safety valve open.

Another type of gas flame sensor is often referred to as the "fire eye." This is actually a device that changes its resistance in the presence of light. When the fire eye is in darkness, it has a very high resistance. When the gas flame is ignited, the light of the flame causes the resistance to change to a low value. Notice that this type of sensor detects the light of a flame and not the heat. This type of control is generally used to sense the presence of the main burner flame instead of the pilot flame. Fire eye detectors are generally used with timers that turn the gas supply off if a flame is not detected within a certain time after a call for heat.

The "flame rod" is another sensor that is generally used to detect the presence of flame at the main burner. The flame rod operates by using the gas flame as a conductor of electricity. A gas flame contains many ionized particles that will conduct electricity in a similar manner to some types of vacuum tubes. When the flame rod is inserted in a flame, a current path exists between the rod and the metal of the burner head itself. As long as there is a flame, there can be a flow of electricity between the rod and the burner head. If the flame should be extinguished, the flow of electricity will stop. Figure 34-3 shows a photograph of a flame rod, a small burner head and an electric spark ignitor.

CONTROL VALVES

The gas control valve is the real heart of the gas heating system. Control valves control the flow of gas to the main burner and the pilot light, if used. Many of them contain an internal pressure regulator, which maintains a constant pressure to the main burner. A simple gas control valve is shown

FIGURE 34-2 Thermocouple and pilot burner

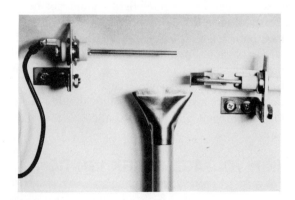

FIGURE 34-3 Flame rod, burner head, and high-voltage ignition electrode

in figure 34-4. This illustration is used to show the basic principle of operation. This type of valve uses a thermocouple to detect the presence of a pilot flame. Notice that a spring is used to close the valve if the thermocouple should stop producing current for the solenoid coil. Also notice that a solenoid coil is used to open the main valve when the thermostat calls for heat. Different valves use different methods of opening the main valve. Some valves use a small electric heater to heat a bimetal strip that opens the main valve. Others use a small heater to cause a metal rod to expand and open the main valve. Regardless of the method used, all control valves perform the same basic function.

The schematic in figure 34-5 shows a basic control circuit for a gas heating system. Notice that the fan and high-limit switch are connected in the 120-volt line ahead of the control transformer. When the thermostat closes, 24 volts AC is applied to the

control valve. This permits the valve to open and supply gas to the main burner. Notice that this control valve uses a thermocouple to sense the presence of a pilot light. If the pilot light should go out, the pilot valve will close and gas flow to the burner will stop.

The schematic in figure 34-6 shows a control circuit that uses a high-voltage spark ignitor. Notice that a fan-limit switch is connected ahead of the 24 volts control transformer. This is the same as the other type of control. In this circuit, however, when the thermostat calls for heat, 24 volts AC is applied to a direct spark ignition control module. When the control module receives a call for heat, it turns on the main control valve and provides about 15,000 volts to the ignition electrode. The module also starts an electronic timer at the same time. When the gas is ignited at the burner head, a current flows from the flame rod to the

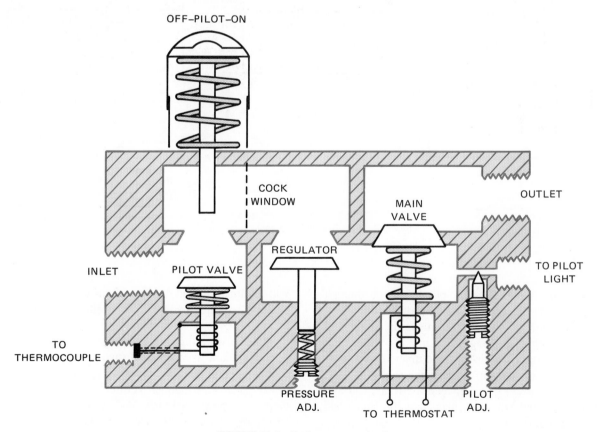

FIGURE 34-4 Basic gas-control valve

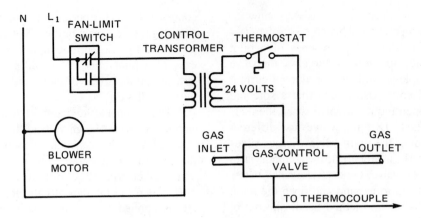

FIGURE 34-5   Basic gas-control system

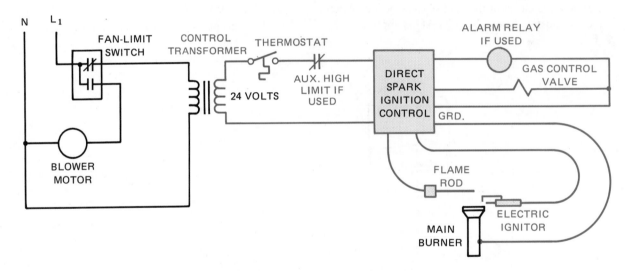

FIGURE 34-6   Electric spark-ignition control

base of the burner. This completes a circuit through the ground wire back to the control module. This flow of current is used to turn off the timer and electric ignitor. As long as a flame is present, and the thermostat calls for heat, the main valve is permitted to remain open. If the flame should go out, however, current flow between the flame rod and burner ground will be broken and the timer and electric ignitor will be started. If a flame is not established in a predetermined time, the main valve

will be turned off and the flow of gas stopped. Some systems are equipped with an alarm relay that is turned on by a solid-state relay when the control module senses an unsafe condition.

If a flame is not established when the thermostat calls for heat, the timer will shut the system down in the same manner. A direct ignition control module is shown in figure 34-7, and a gas control valve used with an electric spark ignition system is shown in figure 34-8.

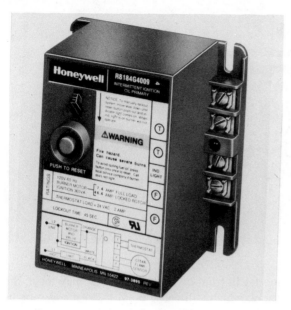

FIGURE 34-7 Direct spark-ignition control module (Courtesy of Honeywell Inc.)

FIGURE 34-8 Main gas valve used with electric ignition (Courtesy of Honeywell Inc.)

REVIEW QUESTIONS

1. What is the purpose of a pilot light?

2. Why is it necessary to be certain that the gas is ignited at the main burner on a call for heat by the thermostat?

3. What is a thermocouple?

4. What is a thermopile?

5. Why must the pilot control valve be reset manually if it should open?

6. Explain how a "fire eye" works.

7. Explain the operation of a "flame rod."

8. What is a common amount of voltage applied to an electric spark ignitor?

9. Why must a ground wire be connected between the direct spark-ignition control module and the burner head?

10. What is the advantage of electric-spark ignition over pilot-light ignition?

UNIT 35
Oil Burner Controls

Some of the controls on an oil-fired heating system are basically the same as the controls on a gas-fired system. The fan and limit controls are very similar and in some cases the same. The major part of an oil-fired control system is the primary control. The primary control's function is to ensure that when the thermostat calls for heat, the flame will be established within a predetermined amount of time. This is to prevent an accumulation of oil vapor in the combustion chamber. If a large amount of oil is formed in the combustion chamber and ignited, an explosion could occur.

IGNITION

A gun-type oil furnace is ignited by an electric arc. Two electrodes are located near the nozzle. When the thermostat calls for heat, the primary control connects the ignition transformer to the 120-volt AC power line. The transformer steps the 120 volts up to 10,000 volts. The 10,000 volts is connected to two electrodes. This causes an arc to be produced between the two electrodes. The air produced by the combustion fan motor causes the arc to be blown in a horseshoe shape as shown in figure 35-1. This arc is used to ignite the oil. The electrodes are adjusted in such a manner that they do not enter into the oil spray produced by the nozzle. Only the horseshoe-shaped arc is permitted to contact the oil spray. If the electrodes are adjusted too far in front of the nozzle, they may touch the spray, which will cause them to burn and soot. If they are adjusted too far behind the nozzle, the arc will not contact the oil spray. This will cause the furnace to start hard and have delayed ignition.

PRIMARY CONTROL

The schematic of a typical primary control is shown in figure 35-2. Notice that this control employs several solid-state components in its operation. These components are:

1. The silicon bilateral switch (SBS)
2. The triac
3. The cadmium sulfide cell (CAD cell)

In this circuit, the gate lead of the SBS has been left disconnected. This permits the SBS to operate very similar to a diac. When the voltage applied to the SBS reaches a high enough level, assume 5 volts for this example, it will turn on and conduct current to the gate of the triac. This will permit the triac to turn on.

CAD CELL

The CAD cell is a device that changes its resistance in accordance with the amount of light it is exposed to. When the CAD cell is in darkness, it

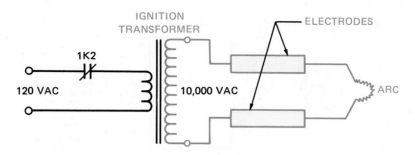

FIGURE 35-1 Electric-arc ignites oil-fired furnace.

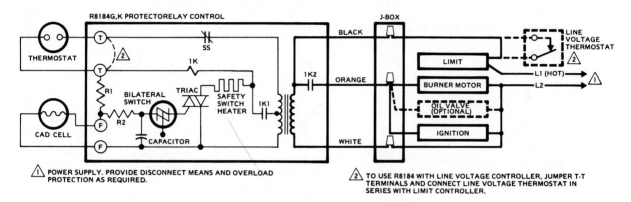

① POWER SUPPLY. PROVIDE DISCONNECT MEANS AND OVERLOAD PROTECTION AS REQUIRED.

② TO USE R8184 WITH LINE VOLTAGE CONTROLLER, JUMPER T-T TERMINALS AND CONNECT LINE VOLTAGE THERMOSTAT IN SERIES WITH LIMIT CONTROLLER.

FIGURE 35-2 Internal schematic and typical hookup for R8184G (Courtesy of Honeywell Inc.)

will have a very high resistance of several hundred thousand ohms. When it is in light, its resistance will decrease to about 50 ohms.

CIRCUIT OPERATION

To help in understanding how this circuit works, it will be shown in different stages of operation. In the circuit shown in figure 35-3, the thermostat has just called for heat. The arrows are used to show the path of current flow through the circuit. The current leaves one side of the step-down transformer and flows through the thermostat contacts. The current then flows through resistor R1. Since the CAD cell is in darkness, it has a very high resistance. This causes most of the voltage to be dropped at the junction point of R1 and R2. Since the voltage at this point is greater than 5 volts, the SBS will turn on and conduct current to the gate of the triac. When the triac turns on, current

is permitted to flow through relay coil 1K, the safety switch heater, the triac, and back to the transformer. Notice that coil 1K is connected in series with the safety switch heater at this time.

Figure 35-4 illustrates the operation of the circuit when relay coil 1K energizes. Notice that both contacts 1K1 and 1K2 are shown closed. When contact 1K2 closes, 120 volts is connected to the burner motor and the ignition transformer. When contact 1K1 closes, a different current path for the relay coil and safety heater is provided to the center tap of the transformer. Relay coil 1K and the safety switch heater are no longer connected in series. Notice that one current path is through the thermostat, and 1K relay coil. The current path through the SBS and triac gate is still provided because the oil flame has not been ignited as yet and the CAD cell is still in darkness.

A second current path is provided through the triac and safety switch heater. If, for some reason, ignition should not occur, current will continue to

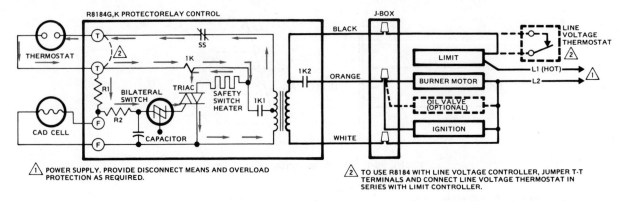

FIGURE 35-3 Internal schematic and typical hookup for R8184G after thermostat has called for heat (Courtesy of Honeywell Inc.)

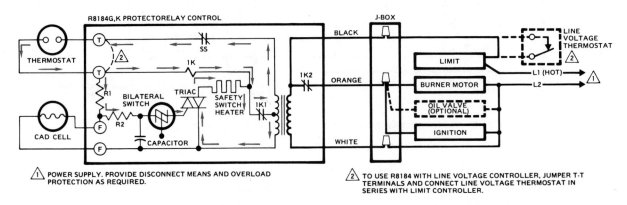

FIGURE 35-4 Internal schematic and typical hookup for R8184G when relay coil 1K energizes (Courtesy of Honeywell Inc.)

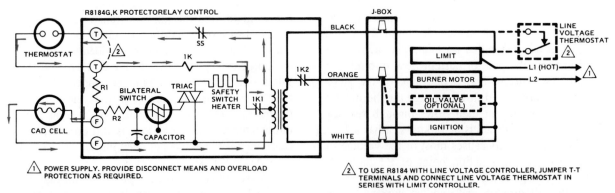

FIGURE 35-5 Internal schematic and typical hookup for R8184G in normal operating condition after ignition (Courtesy of Honeywell Inc.)

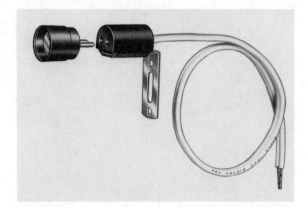

FIGURE 35-6 CAD cell flame detector (Courtesy of Honeywell Inc.)

flow through the triac and safety switch heater. This will eventually cause the bimetal contact SS to open and disconnect the thermostat circuit. If this should happen, it is necessary to manually reset the primary control with the reset button located on the control unit.

In figure 35-5, the circuit is shown in its normal operating condition after ignition. Notice that current is still permitted to flow through the 1K relay coil to keep it energized. The triac, however, has been turned off. When ignition occurs, the CAD cell "sees" the light of the flame. This causes its resistance to drop to a low value. When this happens, the voltage drop at the junction of resistors R1 and R2 becomes very low. Since there is now

FIGURE 35-7 Primary control (Courtesy of Honeywell Inc.)

less than 5 volts, the SBS is turned off and no current is conducted to the gate of the triac. This stops the current flow through the safety switch heater and the circuit continues to operate until the thermostat is satisfied.

A photograph of a CAD cell used as the flame detector in an oil furnace is shown in figure 35-6. A primary control unit for an oil furnace is shown in figure 35-7.

REVIEW QUESTIONS

1. What is the function of the ignition transformer?

2. How much voltage is supplied to the electrodes?

3. Are the electrodes permitted to enter into the oil spray?

4. What does enter into the oil spray to cause ignition?

5. What device is controlled by the operation of the triac?

6. What solid-state device controls the flow of gate current to the triac?

7. Does the CAD cell have a high resistance or low resistance when in the presence of light?

8. How would the circuit operate if the CAD cell should be in the presence of light when the thermostat called for heat?

SECTION 6

Troubleshooting Using Control Schematics

UNIT 36

Room Air Conditioners

In the previous units, basic symbols and rules for reading a schematic diagram have been covered. In actual practice, however, schematics do not always look like the classic textbook examples. Many schematic diagrams use a legend to aid in understanding. A legend is a list that shows a symbol or notation and gives the definition of that symbol or notation. The legend that will be used with the schematics presented in this unit is shown in figure 36-1.

SCHEMATIC #1

The first circuit to be discussed is shown in figure 36-2. First find the major components shown on the schematic: the switch, fan motor, compressor, capacitor, overload, and thermostat. Notice that these components may not be shown exactly as you would expect. Notice the overload symbol, for example. The symbol used is the same as the symbol for an overload heater discussed earlier in the text. There are, however, no overload contacts shown. This schematic is indicating the use of a small, single-phase bimetal overload unit that acts as both heater and overload contact.

Next, find and examine the fan motor. Notice that this motor has several windings that indicate that it is used for multi-speed operation. Notice also that there is no capacitor connected to this motor. The small winding shown separate is the start

winding. Since there is no start or run capacitor shown, this motor is a resistance start induction run. Notice that the white wire is connected from the motor to B terminal of the capacitor and then to one lead of the service chord. This indicates that the white wire is common to the other windings. Now trace the connection of the red, blue, and black wires. The red wire is connected to the LO speed position on the switch; the blue wire is connected to the MED speed position and the black wire is connected to the HI speed position.

Next, examine the compressor. Notice that two windings are shown. Each winding is connected to

Legend

SR	— Start Relay
GND	— Ground
T	— Terminal Bushing
——————	Factory Wiring
– – – –	Field Wiring
⬡	Component Connection

Color Code

Bk or Blk	— Black
Bl or Blu	— Blue
Br or Brn	— Brown
Grn	— Green
Or or Orn	— Orange
R	— Red
Vio	— Violet
Wht	— White
Y-Bk	— Yellow with Black Tracer
Y or Yel	— Yellow

FIGURE 36-1   (Courtesy of Carrier Corp.)

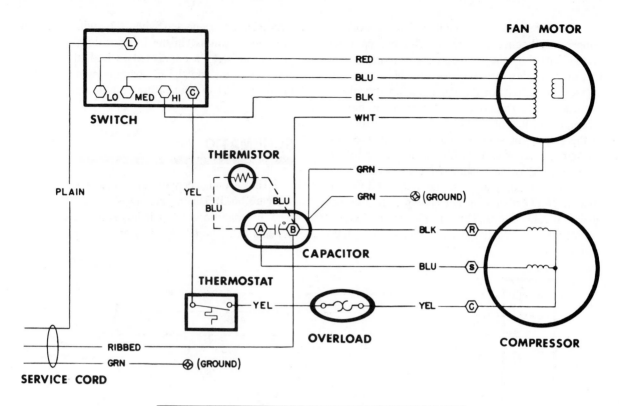

FAN MOTOR

RED
BLU
BLK
WHT

GRN

GRN ⊕ (GROUND)

SWITCH

THERMISTOR

PLAIN

YEL

BLU

BLU

Ⓐ ⊢ B

CAPACITOR

BLK Ⓡ

BLU Ⓢ

THERMOSTAT

YEL

YEL

YEL Ⓒ

COMPRESSOR

OVERLOAD

RIBBED

GRN ⊕ (GROUND)

SERVICE CORD

SWITCH POSITION	CONTACTS MADE
LO	L to C, L to LO
MED	L to C, L to MED
HI	L to C, L to HI
OFF	None

FIGURE 36-2 (Courtesy of Carrier Corp.)

a terminal. One terminal is labeled with an R to represent the run winding. The middle terminal is labeled with an S, which represents the start winding terminal, and the third terminal is labeled with a C, which indicates the common terminal. Trace the common terminal through the overload and thermostat to terminal C on the switch. Notice that the thermostat and overload are connected in series with the compressor. Now trace the run lead of the compressor. Notice that it is connected to B terminal of the capacitor. This shows that the run winding is connected to the common side of the service chord. Now trace the start lead to the A side of the capacitor. Notice that the capacitor is connected in series between the common side of

the service chord and the start winding. The thermistor connected across the capacitor terminals is used to decrease the capacitance connected in series with the compressor after the compressor is in operation. Recall that a thermistor is a temperature-sensitive resistor. This thermistor has a negative temperature coefficient, which means that it will have a very high resistance when it is cool. When its temperature increases, its resistance will decrease. When the compressor is first started, the thermistor is cool because no current has been flowing through it. This causes its resistance to be much greater than the capacitive reactance of the capacitor. The full amount of the capacitor is now connected in series with the start winding.

As current flows through the thermistor, its temperature begins to increase. This causes a decrease in its resistance, which permits more current to flow. As the resistance of the thermistor decreases, the effect of the capacitor on the motor decreases also. The effect is very similar to having a compressor that has both a start and run capacitor in the circuit for starting, and then disconnecting the start capacitor and permitting the motor to operate with the run capacitor only.

The last component to be discussed is the switch. Notice that it is not shown with internal electrical connections. There is a legend at the bottom of the schematic, however, that shows which terminals are connected when the switch is set in different positions. In the LO position, for example, terminal L is connected to both the LO fan speed position and the C position, which permits the thermostat to control the compressor.

SCHEMATIC #2

The second schematic to be discussed is shown in figure 36-3. This schematic is for another room-type air conditioner, but it has some added components. This unit is used to provide heat as well

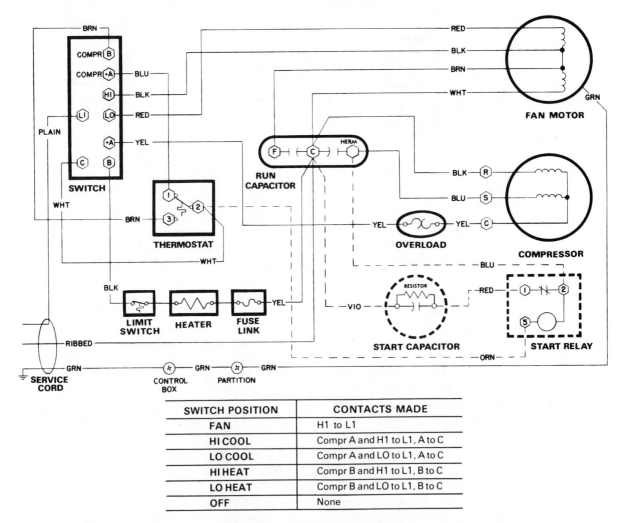

SWITCH POSITION	CONTACTS MADE
FAN	H1 to L1
HI COOL	Compr A and H1 to L1, A to C
LO COOL	Compr A and LO to L1, A to C
HI HEAT	Compr B and H1 to L1, B to C
LO HEAT	Compr B and LO to L1, B to C
OFF	None

FIGURE 36-3 (Courtesy of Carrier Corp.)

SECTION 6 TROUBLESHOOTING USING CONTROL SCHEMATICS

as cooling. An electric resistance heating element is used to provide heat in cool weather. Notice also the addition of the start capacitor and start relay. The thermostat in this circuit is double-acting instead of a single-pole single-throw. This permits the same thermostat to be used for both heating and cooling. Notice also that the run capacitor is different. This capacitor is actually two capacitors contained in the same case. The junction point between the two capacitors is connected to one side of the service chord. The fan motor in this unit is different also. Notice that this fan used a run capacitor connected in series with the start winding of the motor at all timers. This motor is a permanent split-capacitor motor. Notice that this motor has two speeds instead of three.

The start capacitor is connected in parallel with the run capacitor to increase the starting torque of the compressor. The resistor shown connected across the terminals of the start capacitor is a relatively high value of fixed resistance used to discharge the capacitor when it is disconnected from the circuit. Notice the start relay. The start capacitor is connected in series with the normally closed contact. This is a potentional starting relay, which senses the voltage induced in the start winding and opens the contact when the motor reaches about 75% of its full speed.

In this circuit, the switch is the main controller. For example, trace the circuit when the switch is placed in the high cool position. The legend at the bottom of the schematic indicates that when the switch is in the HI cool position, terminals COMPR A and HI are connected to terminal L1, and terminal A is connected to terminal C. When HI is connected to L1, the fan motor will operate in its high speed. When terminal COMPR A is connected to L1, a current path is provided to terminal 1 of the thermostat. Terminal 2 is connected to terminal C of the switch. Since switch terminal A is connected to switch terminal C, power is connected to the compressor motor through the thermostat contact. When the thermostat is connected in this manner, an increase in temperature will cause the thermostat contacts to close and a decrease in temperature will cause them to open.

Now assume that the switch has been set to the low heat position. The switch legend indicates that terminals COMPR B, and LO are connected to L1, and terminal B is connected to terminal C. When LO is connected to L1, the fan motor operates in the low speed position. When terminal COMPR B is connected to L1, power is provided to terminal 3 of the thermostat. Terminal 2 is connected to terminal C of the switch. Since switch terminal B is connected to the resistance heater through the high limit switch and fuse, the thermostat controls the operation of the heater. When the thermostat is connected in this manner, a decrease in temperature will cause the thermostat contacts to close and an increase in temperature will cause them to open.

REVIEW QUESTIONS

1. What is a legend?

2. Refer to figure 36-2. What would be the action of this circuit if the overload relay should burn open?

3. What purpose does the thermistor connected in parallel with the capacitor serve?

4. In figure 36-2, what switch connections are made when the switch is in the HI position?

5. In figure 36-3, why is the thermostat switch shown as a single-pole double-throw?

6. In figure 36-3, what do the dashed lines showing connection between the start capacitor and start relay to other parts of the circuit mean?

7. In figure 36-3, what color wire is connected between terminal 2 of the thermostat and terminal C of the switch?

8. What color wire is connected between terminal 2 of the thermostat and the start relay?

9. In figure 36-3, if no continuity is shown when one lead of an ohmmeter is connected to switch terminal A and the other is connected to terminal C of the compressor, what does it mean?

10. In figure 36-3, to what two points should the terminals of an ohmmeter be connected to check the continuity of the resistance heater circuit?

UNIT 37

A Commercial Air-Conditioning Unit

In this unit, a commercial air-conditioning system will be discussed. The legend for this schematic is shown in figure 37-1. The schematic to be discussed is shown in figure 37-2. Notice that this control system contains several devices not normally found in a residential system. The compressor, for example, is operated by a three-phase squirrel cage induction motor. It can be seen that the motor is three phase by the wye connection of the stator winding. It can be determined that the motor is a squirrel cage because it has no external resistors that would be used for the rotor circuit of a wound rotor induction motor. There is also no DC circuit that would be required to excite the rotor of a three-phase synchronous motor.

The condenser fan motor is a single-phase permanent split-capacitor motor. Notice that the condenser fan motor is connected in parallel with two lines of the compressor. When contactor C energizes, both C contacts close and connect both the compressor and condenser fan motors to the line.

The crankcase heater is shown directly below the condenser fan motor, and is connected to terminal 21 and 23. Notice the crankcase heater is energized at all times. As long as power is connected to the circuit, the crankcase heater will be energized.

The control transformer contains two primary windings and two secondary windings. This transformer can be connected to permit a 460- or 230-volt connection to the primary, and the secondary can provide 230 or 115 volts. In the circuit shown,

the primary winding is connected in series, which permits 460 volts to be connected to it. The secondary winding is also connected in series, which provides an output voltage of 230 volts.

The 230-volt circuit is used to operate a short-cycle timer circuit. This is the same circuit that was discussed in unit 30.

The 24-volt circuit is shown at the bottom of the schematic. Notice that only the secondary of the transformer is shown. This is indicating that its power can be derived from almost anywhere. The primary of this transformer could be connected to a 120-volt circuit inside the building. This circuit contains the high- and low-pressure switches. If one of them should open, it will have the same effect as opening the thermostat.

Notice that the indoor fan relay (IFR) is shown, but the fan motor is not. In a commercial location, there may actually be several fans operated by the IFR relay. In practice, the IFR relay may be used to control the coils of other relays, which connect the fan motors to the line.

The thermostat is a single-pole single-throw contact. The resistor shown connected around the thermostat contact represents the heat anticipator. A switch is also provided that will permit the indoor fan to be operated automatically or manually.

The last item shows the component arrangement. This is used to aid the service technician in locating the different control components in the system.

LEGEND

C	— Contactor		SC	— Start Capacitor
CC	— Cooling Compensator		SR	— Start Relay
CH	— Crankcase Heater		ST	— Start Thermistor
Comp			TC	— Thermostat, Cooling
or	— Compressor		TD	— Time Delay
Compr			Therm	— Thermostat
CPCS	— Compressor Protection Control System		TM	— Timer Motor
CR	— Control Relay		Tran	
CT	— Current Transformer		or	— Transformer
FC	— Fan Capacitor		Trans	
FM	— Fan Motor		⬡	Component Connection (Marked)
FS	— Fan Switch		o	Component Connection (Unmarked)
FT	— Fan Thermostat			Field Splice
HC	— Heating Control			Splice
HPS	— High Pressure Switch			
HR	— Holding Relay		→	Plug
IFM	— Indoor Fan Motor			Receptacle
IFR	— Indoor Fan Relay			
IP	— Internal Protector			Factory Wiring
LPS	— Low Pressure Switch			Field Power Wiring
OL	— Overload			Field Ground Wire
QT	— Quad Terminal			Field Control Wiring (NEC, Class II)
R	— Resistor			Alternate Wiring
RC	— Run Capacitor			Indicates common potential
Recep	— Receptacle			(Does not represent wire)
Res	— Bleed Resistor			

FIGURE 37-1   (Courtesy of Carrier Corp.)

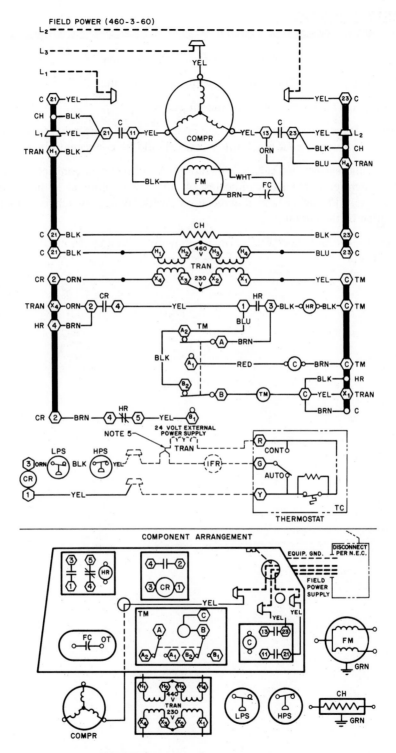

FIGURE 37-2 (Courtesy of Carrier Corp.)

REVIEW QUESTIONS

1. What does the term CC mean if seen on a control schematic?

2. What does the term CPCS mean if seen on a schematic?

Refer to figure 37-2 for the following questions.

3. If it was desired to change the voltage controlling the short-cycle timer from 230 volts to 115 volts, what transformer leads should be connected together?

4. Assume the system has stopped operation. A voltmeter is connected across the LPS switch terminals and it indicates 24 volts. The voltmeter is then connected across the HPS switch and it indicates 0 volts. Which switch is stopping the operation of the circuit?

5. When the system is operating normally, how much voltage should be seen across the CR relay coil?

UNIT 38

Heat-Pump Controls

A heat pump is a device that provides both heating and air conditioning within the same unit. In the cooling cycle, the outside heat exchange unit is used as the condenser and the inside heat exchanger is used as the evaporator. When the heat pump is used for heating, the reversing valve reverses the flow of refrigerant in the system and the outside heat exchanger becomes the evaporator. The inside heat exchanger becomes the condenser. Heat pumps also contain some type of back-up heating system that is used when the outside temperature is too low to make heat transfer efficient. The most common type of back-up heat is electric-resistance heat.

Heat pumps contain other control devices that are generally used only with heat-pump equipment, such as double-acting thermostats, sequence relays, and defrost timers.

DOUBLE-ACTING THERMOSTATS

The double-acting thermostat is a thermostat that contains two separate mercury contacts. It is similar to the programmable thermostat except that the two mercury contacts cannot be set independently of each other. The mercury contacts of the double-acting thermostat are so arranged that one contact will make connection slightly ahead of the other. For example, assume the heat pump is being used in the heating mode. Now assume that the

temperature drops. One of the contacts will make connection first. This contact turns on the compressor and heat is provided to the living area. If the compressor can provide enough heat to raise the temperature to the desired level, the second mercury contact does not make connection. If the compressor cannot provide the heat needed, the second mercury contact will close and turn on the electric-resistance heating elements to provide extra heat to the living area.

THE SEQUENCE TIMER

The sequence timer is an on-delay timer used to connect the heating elements to the line in stages instead of all at once. Most sequence timers contain two or three contacts and are operated by a small heating element that heats a bimetal strip. When the bimetal strip becomes hot enough, it snaps from one position to another and closes the two contacts. A basic drawing of this type of timer is shown in figure 38-1.

DEFROST TIMER

When the heat pump is used in the heating mode of operation, it removes heat from the air and delivers it inside the living area. This means that the outside heat exchanger is being used as the

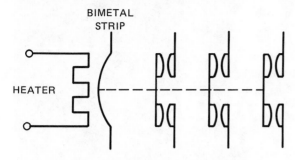

FIGURE 38-1 Sequence relay

evaporator and cold refrigerant is circulated through it. Any moisture in the air can cause frost to form on the coil and reduce the airflow through it. This will reduce the efficiency of the unit. For this reason, it is generally necessary to defrost the outside heat exchanger. Defrosting is done by disconnecting the condenser or outside fan motor and reversing the flow of refrigerant through the coil. This causes the unit to temporarily become an air conditioner and warm refrigerant is circulated through the coil.

Before the defrost cycle can be activated, two separate control conditions must exist. The defrost thermostat, located on the outside heat exchanger, must be closed; and the defrost timer must permit the defrost cycle to begin. A schematic diagram of a basic defrost control circuit is shown in figure 38-2. Notice that the defrost timer is connected in parallel with the compressor. This means that the timer can operate only when the compressor is in operation. Notice also that the defrost cycle energizes the reversing valve solenoid. This means that

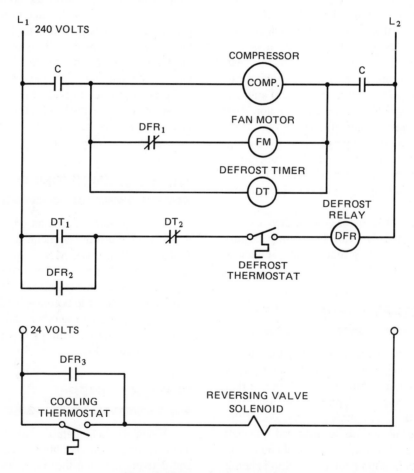

FIGURE 38-2 Defrost cycle circuit

this unit is in the heating mode when the solenoid is de-energized.

Notice the defrost timer contains two contacts, DT1 and DT2. DT1 is normally open and DT2 is normally closed. The defrost relay (DFR) contains three contacts. DFR1 is normally closed and is connected in series with the outside fan motor. DFR2 is normally open and is connected in parallel with contact DT1. Contact DFR3 is normally open and is connected to the reversing valve solenoid.

The schematic shown in figure 38-3 illustrates the condition of the circuit when the defrost cycle first begins. Notice that the defrost timer (DT) has caused contact DT1 to close, but contact DT2 has not opened. The contacts of the defrost timer are operated by two separate cams. The cams are so

arranged that contact DT1 will close before DT2 opens.

The schematic in figure 38-4 illustrates the condition of the circuit immediately after the defrost relay has energized. Notice that all DFR contacts have changed position. DFR1 contact is now open and the outside fan motor has been disconnected from the circuit. DFR2 contact is closed and is used as a holding contact around contact DT1. DFR3 contact is closed and provides current to the reversing valve solenoid to reverse the flow of refrigerant in the system.

The schematic shown in figure 38-5 illustrates the condition of the circuit after contact DT1 reopens. Notice that contact DFR2 maintains a current flow path around the now open DT1 contact

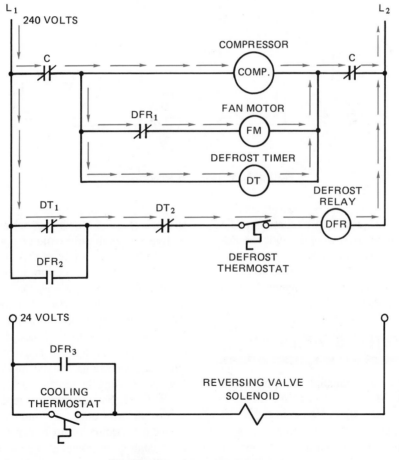

FIGURE 38-3 Initial circuit operation

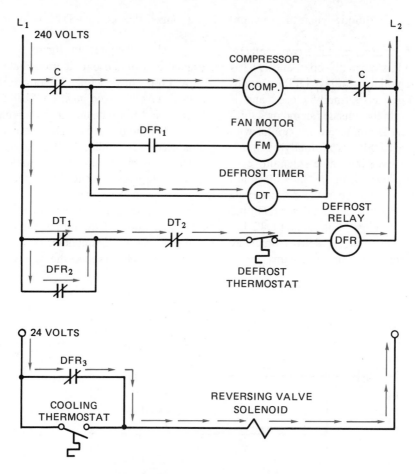

FIGURE 38-4 Defrost relay energizes.

and the defrost cycle is permitted to continue. The unit will remain in the defrost cycle until the defrost thermostat is satisfied and opens the circuit, or the defrost timer causes the DT2 contact to open. When this occurs, the system will change back to its original condition shown in figure 38-2.

THE FULL SYSTEM SCHEMATIC

A schematic for a residential heat pump unit is shown in figure 38-6. The legend for the schematic is shown in figure 38-7. Notice in figure 38-6 that the schematic is divided into three main sections. One section shows the outdoor compressor controls. The second section shows the indoor re-

sistance heat and blower-fan controls, and the third section shows the low-voltage controls.

To begin the study of this control system, locate the low-voltage section of the schematic. It is divided into three sections. One section is located directly below the blower fan motor. Notice the 24-volt transformer used to provide needed power. Now locate the terminal board directly to the left of the control transformer. The terminal board shows terminal connections marked inside hexigon-shaped figures. Starting at the top and going down they are R, G, O, Y, etc. Now locate the second control section directly under the outdoor unit schematic. Notice the terminal board contains some of the same letter connection points as the other terminal board. Now locate the thermostat. Notice the thermostat

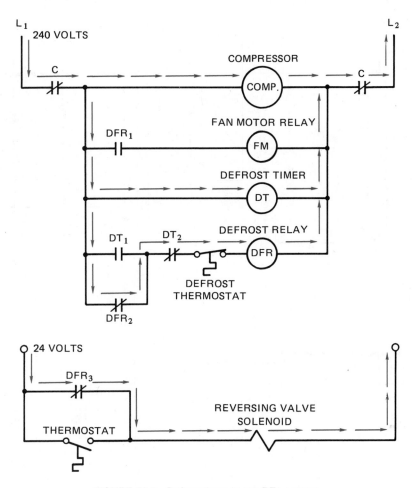

FIGURE 38-5 Defrost timer opens DT1 contact.

contains terminal markings that are the same as the other two boards. These terminal markings are used to aid in tracing the circuit. For example, locate the terminal marked Y on the thermostat. Now, locate the terminal marked Y on the board closest to the control transformer. Finally, locate the terminal marked Y on the terminal board located under the outdoor unit. If the wires are traced, it will be seen that all of the terminals marked Y are connected together. This is true for all the other terminals that are marked with the same letter. Terminal markings are often used to help simplify a schematic by removing some to the connecting wires. The circuit shown in figure 38-8 is very similar to the schematic in figure 38-6 except that

the terminal markings are used instead of connecting wires.

Notice the use of the double-acting thermostat in the schematic shown in figure 38-6. The thermostat is so constructed that contact H1 will close before H2. Also, contact CO will close before C1. Also notice that the thermostat has an emergency heat position that permits the switch to override the thermostat and the electric resistance heaters to be connected in the circuit on a call for heat. Notice also that this control uses an outdoor thermostat (ODT). The ODT senses the outside temperature and permits sequence relays 1 and 2 to operate only if the outside temperature is below a certain level.

Locate the three sequence relay coils, SEQ1,

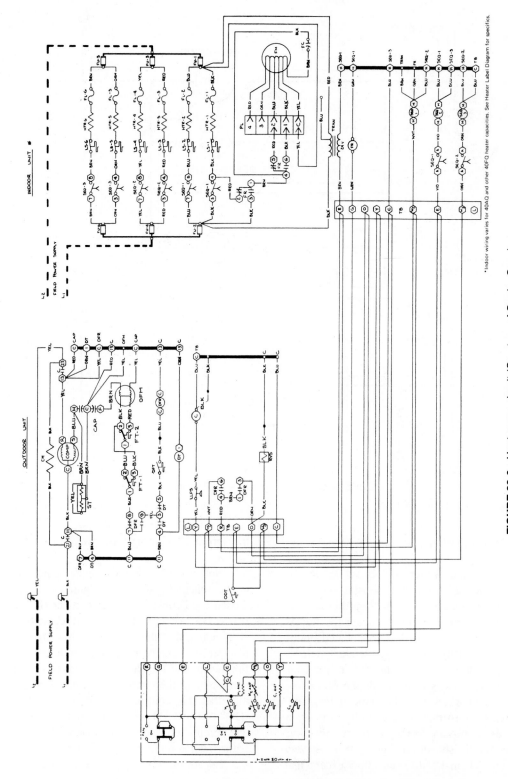

FIGURE 38-6 Heat-pump circuit (Courtesy of Carrier Corp.)

SECTION 6   TROUBLESHOOTING USING CONTROL SCHEMATICS

LEGEND

Ant.	— Anticipator		LS	— Limit Switch
C	— Contactor		ODT	— Outdoor Thermostat
Co	— First-Stage Cooling Thermostat		OFM	— Outdoor Fan Motor
C1	— Second-Stage Cooling Thermostat		PI	— Plug
Cap.	— Capacitor		QT	— Quad Terminal
CH	— Crankcase Heater		RC	— Run Capacitor
Comp	— Compressor		RVS	— Reversing Valve Solenoid
DFR	— Defrost Relay		SC	— Start Capacitor
DFT	— Defrost Thermostat		Seq	— Sequencer
DT	— Defrost Timer		SR	— Start Relay
Em Ht	— Emergency Heat		ST	— Start Thermistor
EHR	— Emergency Heat Relay		TB	— Terminal Board
FC	— Fan Capacitor		Tran	— Transformer
FL	— Fuse Link			
FT	— Fan Thermostat		▬▬▬	Indicates Common Potential Only. Does Not Represent Wire.
Fu	— Fuse			
H1	— First-Stage Heating Thermostat		———	Factory Wiring
H2	— Second-Stage Heating Thermostat		– – –	Field Power Wiring
Htr	— Heater		– – –	Field Control Wiring
IFM or FM	— Indoor Fan Motor		⌐Я⌐	Field Splice
IFR or FR	— Indoor Fan Relay		⬡	Component Connection (Marked)
			◯	Component Connection (Unmarked)
LLPS	— Liquid Line Pressure Switch		●	Junction

FIGURE 38-7 (Courtesy of Carrier Corp.)

SEQ2, and SEQ3. Trace the operation of the circuit when sequence relay 1 is energized. The two SEQ1 contacts located in the resistance heat section close and connect the heating elements to the line. A third SEQ1 contact is connected in series with SEQ2 timer. When this contact closes, SEQ2 timer can begin operation provided the outdoor thermostat is closed. When this timer completes its time sequence, all SEQ2 contacts close. The two SEQ2 contacts located in the resistance heat section connect the second bank of resistance heaters to the line. The third SEQ2 contact permits current to flow to SEQ3 timer. At the end of its time cycle, the two SEQ3 contacts located in the resistance heater section close and connect the third bank of heaters to the line.

Locate the blower fan motor. Notice that this is a multi-speed fan motor. Only two of the speeds are used, however. High speed is used when the

fan control relay is energized by the thermostat. Notice that when the normally open FR contacts close, high speed is connected to the line. Also notice that the normally closed FR contacts are connected to the first SEQ1 contact. When SEQ1 contact closes, the second fan speed is connected to the line.

Now locate the defrost timer and defrost relay. Trace the action of the circuit as described earlier in this unit. Notice that the DFR contact located between terminals 4 and 6 is used to override the outdoor thermostat. This permits the resistance heaters to be used during the defrost cycle, preventing cold air from being blown in the living areas during the time the unit is operating as an air conditioner.

Now locate the two outside fan-speed control thermostats labeled FT1 and FT2. Notice that when FT1 is in the position shown it permits FT2 to op-

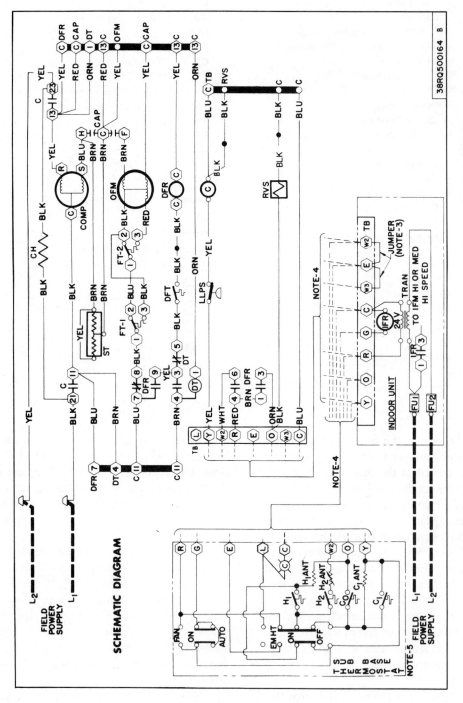

FIGURE 38-8 Terminal identification is used to simplify schematic. (Courtesy of Carrier Corp.)

erate the fan in either high speed or low speed. When FT1 changes position and makes connection between terminals 1 and 3, it connects the fan in the high-speed position and FT2 has no control over the speed.

REVIEW QUESTIONS

1. What is the purpose of terminal markings?

2. What two control components must be in a closed position before a heat pump is permitted to go into the defrost cycle?

3. The thermostat shows a small pilot light connected between terminals L and C. What condition of the thermostat turns this light on?

4. What is the purpose of the outdoor thermostat?

5. What is the operating voltage of the reversing valve solenoid?

UNIT 39

Packaged Units: Electric Air Conditioning and Gas Heating

The schematic shown in figure 39-1 is for a unit that contains both electric air conditioning and gas heating in the same package. This drawing shows both a connection diagram and a schematic diagram of this unit.

This diagram shows mainly the heating and blower controls. At the bottom of the schematic diagram is a component labelled *Condensing Unit*. This is the only reference to the air-conditioning compressor and condenser fan on this schematic. This is not uncommon for a packaged unit.

THE COOLING CYCLE

The thermostat shows 4 terminal connections. The terminal labelled "R" is connected to one side of the 24-volt control transformer. When the thermostat is in the cooling position, an increase in temperature will cause terminal "R" to make connection with terminals "G" and "Y." When power is applied to terminal "Y," a circuit is completed to the condensing unit. The other side of the condensing unit is connected to terminal "C," which completes the circuit back to the control transformer. This starts the air-conditioning compressor and condenser fan.

Terminal "G" of the thermostat is connected to the blower relay coil (BLR). When BLR coil energizes, both BLR contacts change position. The normally closed contact opens and prevents the possibility that power can be applied to the low-speed terminal of the blower fan motor. The normally open contact closes and connects power to the high-speed terminal of the motor. Notice that the indoor blower fan operates in the high-speed position when the air-conditioning unit is started.

THE HEATING CYCLE

When the thermostat is in the heating position, a decrease of temperature will cause the thermostat to make connection between terminals "R" and "W." This permits a circuit to be completed through the automatic gas valve (AGV). When the AGV is energized, gas is permitted to flow to the main burner where it is ignited by the pilot light. Two high-limit contacts are connected in series with the automatic gas valve. One is labelled *Auxiliary Limit*, and the other is labelled *Main Limit*. The wiring diagram shows the main limit to be located in the fan-limit switch. The auxiliary limit switch is in a separate location. Both of these switches are normally closed and are shown to be temperature activated. The schematic also shows that an increase in temperature will cause them to open. Since both are connected in series with the AGV, the circuit will be broken to the valve if either one opens.

In the heating cycle, the indoor blower fan is controlled by the fan switch. The fan switch is temperature activated. After the gas burner has been

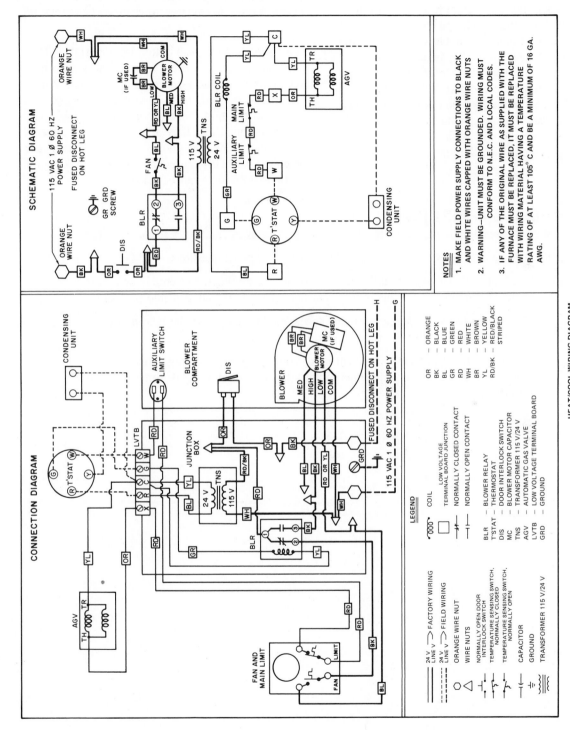

FIGURE 39-1   Schematic for an air-conditioning and heating package unit (Courtesy of Borg-Warner Central Environmental Systems Inc.)

turned on, the temperature of the furnace increases. When the temperature has risen to a high enough level, the fan switch will close and connect the low-speed terminal of the blower motor to the power line. Notice that the fan switch is connected in series with the normally closed BLR contact. When BLR relay is de-energized, the fan switch is permitted to control the operation of the blower fan. The blower fan relay permits the fan to operate in low speed when the unit is in the heating cycle, and in high speed when the unit is in the cooling cycle.

THE DOOR INTERLOCK SWITCH

The door interlock is shown on the schematic as a normally open push button labelled (DIS). The function of this switch is to permit the unit to operate only when the furnace door is closed. When the door is opened, the 120-volt power supply is broken to the unit. Most door interlock switches are so designed that they are actually a two-position switch. When the door is open, the switch can be pulled out. This causes the switch to make connection so the unit can be serviced.

REVIEW QUESTIONS

Refer to the schematic shown in figure 39-1 for the following questions.

1. The unit will operate normally in the cooling cycle. When the unit is switched to the heating cycle, the gas burner will not ignite. List four possible problems.

2. The blower fan will operate normally in the cooling cycle. In the heating cycle, however, the fan will not operate. List three possible problems.

3. The unit will not operate in the heating or cooling cycle. A voltage check shows that there is no low voltage for operation of the control circuit. List three possible problems.

SECTION 7

Ice Makers

UNIT 40
Household ice makers

Ice makers can be divided into two major categories, household and commercial. Unlike commercial units, household ice makers do not recirculate water. They fill a tray or mold and the water is allowed to freeze. Various methods are used to sense when the water has been frozen and to eject the ice from the tray.

Commercial ice makers generally recirculate the water during the freeze cycle. The one reason for this is that pure water freezes faster than water containing impurities and minerals. The ice formed is more pure and clearer in color. This does not apply to flaker-type machines, however. Flaker or crushed-ice machines use an auger to scrape ice off an evaporator after the water has been frozen.

Some cube-type machines freeze water in the shape of the cube, and others freeze water as a slab. The slab-type machines use a grid of cutter wires to cut the frozen slab into cubes.

HOUSEHOLD ICE MAKERS

One of the most widely used household ice makers is the compact, figure 40-1. Although a newer model has been introduced, many of these original units are still in operation. The basic operation of this unit is as follows:

1. An electric solenoid valve, figure 40-2, turns on and fills the tray or mold with water. The valve contains a flow washer which meters the amount of water. The washer is designed to work with pressures that range between 15 and 100 PSI. The length of time the water is permitted to flow is controlled by a cam operated by a small electric motor. The time can be adjusted by moving the water solenoid switch closer to or farther away from the cam. The amount of water needed to fill the mold is approximately 135 cc or 4 oz. It should be noted that insufficient water causes the thermostat to cool too quickly, causing the ice maker to eject hollow cubes.

2. A thermostat senses when the water is frozen. It is mounted directly on the mold by a spring clip. The thermostat controls the start of the ejection and refill cycle.

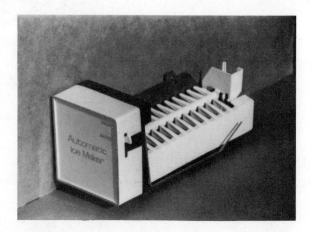

FIGURE 40-1   Early model of a compact ice maker

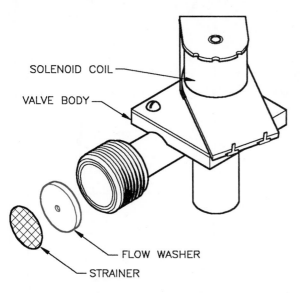

FIGURE 40-2   Water valve

3. When the thermostat contact closes, it turns on the mold heater and motor. The motor operates the timing cam and ejector blades. The ice maker is so designed that the ejector blades can stall against the ice cubes without causing harm to the motor or mechanical parts. When the heater has warmed the mold sufficiently, the ice cubes are pushed out by the ejector blades.

4. During the ejection cycle, the shutoff arm rises and lowers. The shutoff arm senses the height of ice in the holding bin. If the bin is not full, the arm returns to its original position and the ice maker is permitted to eject ice cubes again after they have been frozen. If the holding bin is full, however, the arm cannot return to its normal position and the next ejection cycle cannot begin. The ice maker can be manually turned off by raising the shutoff arm above its normal range of travel.

The ice maker will normally permit the ejector blades to make two revolutions before the thermostat reopens its contact and permits the process to stop at the end of the cycle. If the ejector blades make only one revolution, the ice cubes will be left on top of the blades instead of being dumped into the holding bin. This is not a problem, however, because the cubes will be dumped at the beginning of the next ejection cycle. Near the end of the cycle the mold is refilled with water.

OPERATION OF THE CIRCUIT

The basic circuit for the compact ice maker is shown in figure 40-3. The circuit is shown during the freeze cycle. It is assumed that the mold has

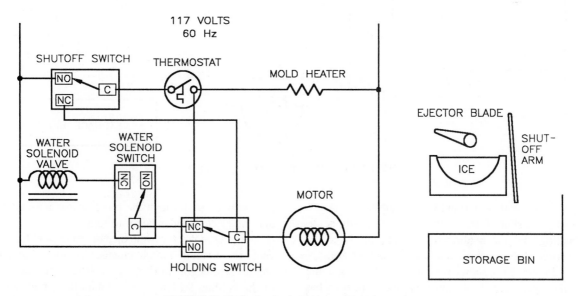

FIGURE 40-3   Basic circuit for the compact ice maker

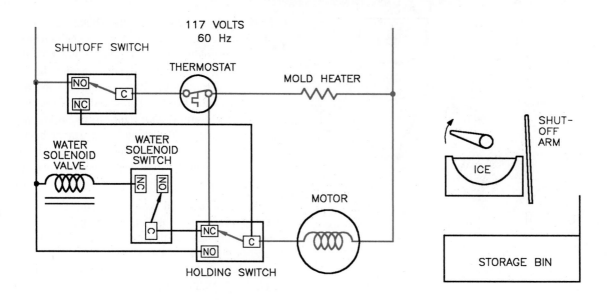

FIGURE 40-4   Beginning of the ejection cycle

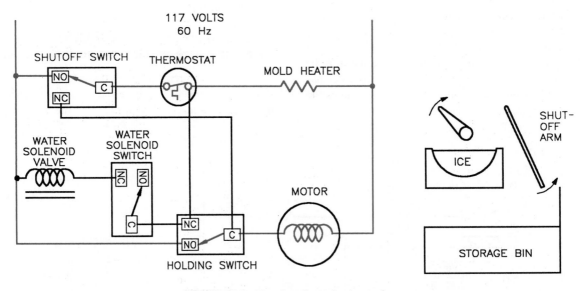

FIGURE 40-5   The shutoff arm begins to rise.

been filled with water and the thermostat contact is open. The shutoff arm is in its normal position, indicating that the holding bin is not full. Note the position of the ejector blade and the shutoff arm.

Figure 40-4 shows the circuit at the beginning of the ejection cycle. At this time, the thermostat has cooled sufficiently for its contact to close. A current path now exists through the mold heater and motor, and the ejector blades begin to turn.

As the motor turns the ejector blades and timing cam, the holding switch changes position and the shutoff arm begins to rise, figure 40-5. The function of the holding switch is to maintain the circuit until the cam returns to the freeze, or off, position.

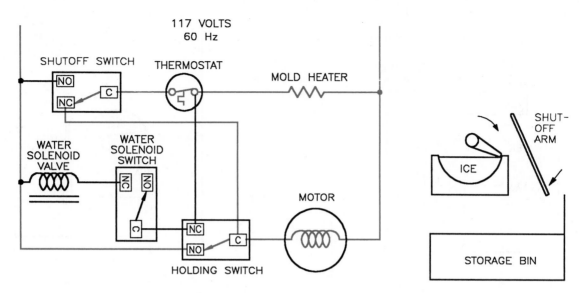

FIGURE 40-6 The shutoff switch changes position.

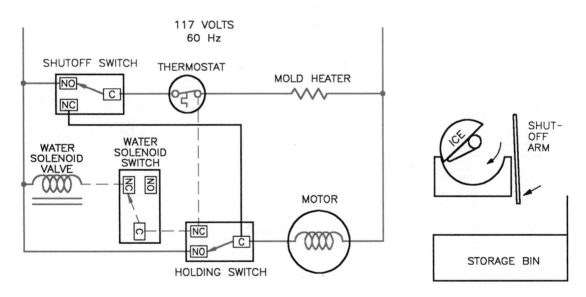

FIGURE 40-7 The timing cam closes the water solenoid switch.

In figure 40-6, the timing cam causes the shutoff arm to rise and fall, making the shutoff switch change position. When the ejector blades reach the ice in the mold, the motor will stall until the ice cubes are thawed loose by the mold heater. Notice that the circuit to both the heater and motor has been maintained by the holding switch. Note that it is possible for the thermostat to open its contact at any point in this process. If this should occur, power is turned off to the mold heater but maintained to the motor by the holding switch.

Near the completion of the first revolution of the ejector blades, the timing cam closes the water solenoid switch, figure 40-7. Although the water solenoid switch is now closed, current cannot flow through the coil. As long as the thermostat contact

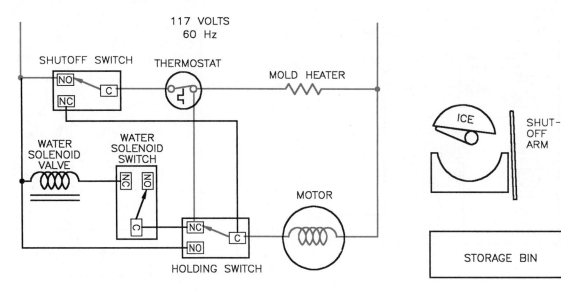

FIGURE 40-8 End of the first revolution

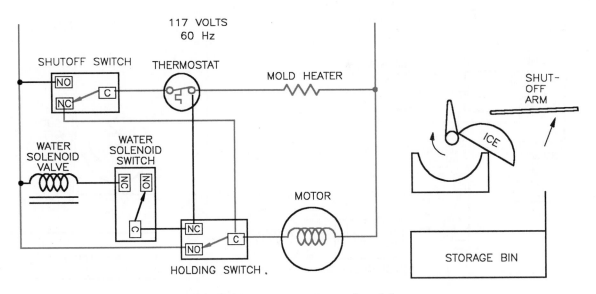

FIGURE 40-9 Beginning of second revolution

is closed, the same voltage potential is applied to both sides of the water solenoid coil. Since there is no potential difference across the coil, no current can flow and the water valve does not open to permit water flow into the mold.

At the end of the first revolution, figure 40-8, the shutoff arm and ejector blades have returned to their normal position and the timing cam has reset all cam-operated switches back to their normal position. Notice, however, that the thermostat contact has remained in the closed position, permitting the second revolution to begin.

After the timing cam has rotated a few degrees, the holding switch again closes to maintain a current path to the motor and mold heater, figure 40-9. The shutoff arm raises and changes the position of the

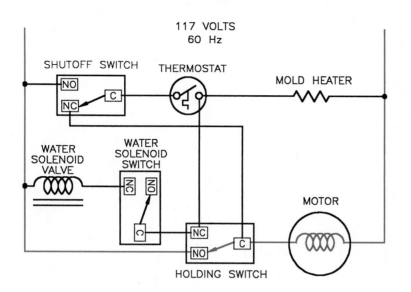

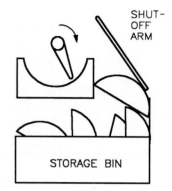

FIGURE 40-10 Middle of second revolution

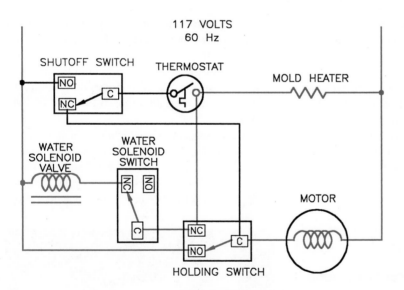

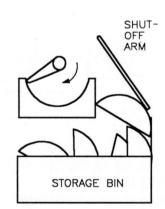

FIGURE 40-11 End of second revolution

shutoff switch. The continued rotation of the ejector blades dumps the ice into the holding bin.

During the second revolution, the increased temperature from the mold heater causes the thermostat contact to reopen, which deenergizes the heater, figure 40-10. The holding contact, however, provides a continued current path to the motor. If the storage bin is full, the shutoff arm will not return to its normal position and the shutoff switch will not be reset.

Near the completion of the second revolution, figure 40-11, the timing cam again closes the water solenoid switch. A current path now exists through the solenoid coil and the mold heater. Although the solenoid coil and mold heater are now connected in series, the impedance of the solenoid coil is much

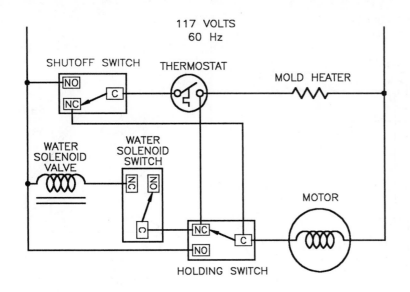

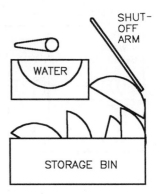

FIGURE 40-12 End of cycle

higher than that of the heater. This permits most of the voltage, about 105 to 110 volts, to be applied across the coil, causing it to energize and open the water valve.

The cycle ends when the timing cam reopens the water solenoid and holding switch, figure 40-12. If the storage bin is full as shown in this illustration, a new ejection and refill cycle cannot begin until sufficient ice has been removed from the storage bin to permit the shutoff arm to return to its normal position.

THE NEW MODEL COMPACT ICE MAKER

Although the new model compact ice maker, figure 40-13, is very similar in design and operating principle to the original version just discussed, there are some significant differences. Some of these differences are listed below:

1. The ejector blades on the newer model stop at a different position, as shown in figure 40-14. Also shown in figure 40-14 is the position of the ejector blades when different actions occur during the ejection cycle.

2. The ejector blades make only one revolution instead of two during the ejection cycle.

3. Most of the new models have an external water level adjustment knob, figure 40-13. Turning the knob moves a set of contacts in relation to its contact ring, permitting the fill time to be longer or shorter.

4. On the original model compact ice maker, the gear located on the front of the unit could be turned manually to advance the ice maker through different parts of the cycle. This gear should **NEVER** be turned on the newer model.

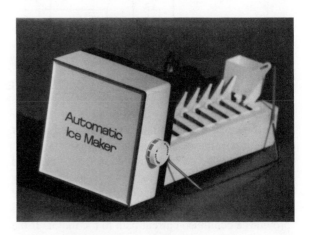

FIGURE 40-13 New compact ice maker

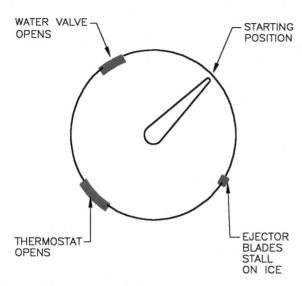

FIGURE 40-14 Ejector blade positions on new model compact ice maker

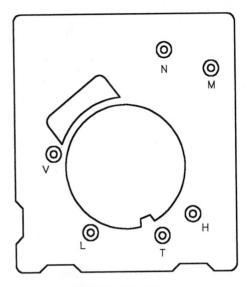

FIGURE 40-16 Test points

To do so will cause damage to the ice maker. The front gears of both the original and newer compact ice makers are shown in figure 40-15.

5. The new model compact provides test points on the plate located behind the front cover, figure 40-16. It is possible to test different parts of the electrical circuit using a voltmeter and ohmmeter. The letters indicate the following test points:

- N = Neutral side of the line
- L = L1 (HOT) side of the line

FIGURE 40-15 Front gears of the original (left) and newer model (right) compact ice makers

- M = Motor connection
- H = Heater connection
- T = Thermostat connection
- V = Water valve connection

6. Probably the greatest difference lies in the electrical circuit itself. In this model, copper strips are laminated on an insulated plate located on the back side of the drive gear. When the motor turns, it turns these copper strips also. Contacts ride against these copper strips and make or break connection to operate the circuit. A diagram of the copper strips and contacts is shown in figure 40-17.

The basic electrical circuit for this unit is shown in figure 40-18. Please note that the contact points A, B, C, and D correspond to the contacts shown in figure 40-17. At this point, connection is made between contacts B and C.

When the thermostat reaches approximately 17°F, its contact will close and produce a current path to both the motor and heater as shown in figure 40-19. The motor begins to turn both the ejector blades and the copper strips located on the back of the main gear. At some point, contact between points B and C is broken and contact between points

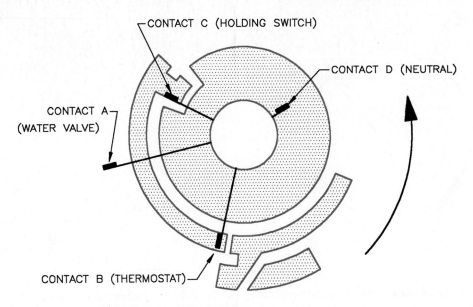

FIGURE 40-17 Rotary switch located on back of drive gear

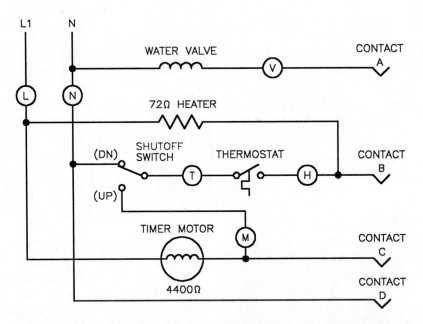

FIGURE 40-18 Basic schematic diagram for new model of whirlpool compact ice maker

C and D is made, as shown in figure 40-20. The ejector blades then stall against the ice. A current path is maintained to the motor between points C and D and a current path is maintained to the heater by the closed thermostat contact.

After the surface of the ice has been thawed by the heater, the ejector blades will begin to turn again. After the ejector blades have rotated approximately 180°, the thermostat contact opens, figure 40-14. As the blades continue to turn, the shutoff

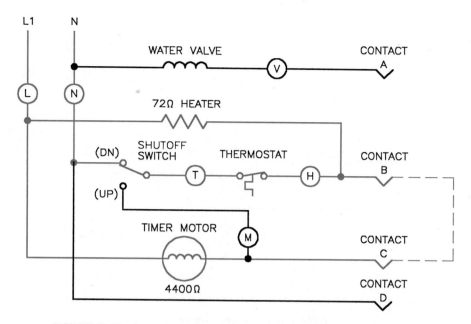

FIGURE 40-19 A current path is provided through the heater and timer motor.

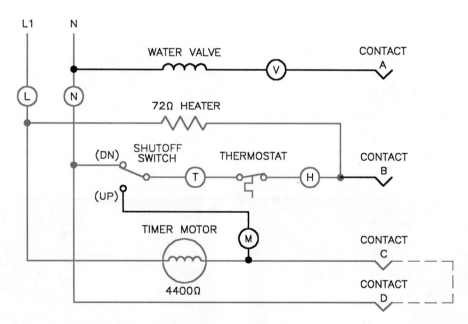

FIGURE 40-20 The holding contact maintains the timer motor circuit.

arm rises and lowers and the copper strips advance until connection is made between contacts A and B, figure 40-21. This provides a current path through the mold heater to the water solenoid valve. Since the coil of the solenoid has a much higher impedance than the mold heater, most of the line voltage will be dropped across the valve, causing it to open and refill the mold. The ejector blades will continue to

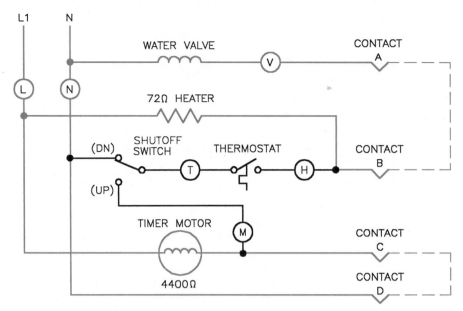

FIGURE 40-21 Water valve energizes

turn until they reach the end of the cycle and the circuit returns to its original condition as shown in figure 40-18.

FLEX TRAY ICE MAKERS

Another type of household ice maker is known as the *flex tray,* figure 40-22a and 40-22b. This unit differs from the compact ice maker in several ways. The flex tray ice maker fills a tray with water, and then after some length of time, it turns the tray to dump the cubes into a storage bin, figure 40-23. At a point during the ejection cycle, a tab located on one side of the rear of the tray contacts a stationary stop. The front of the tray continues to turn, causing the tray to flex or bend at about a 20° angle. This flexing action causes the cubes to dump into the stor-

FIGURE 40-22A Side view of flex tray ice maker

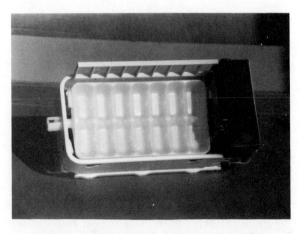

FIGURE 40-22B Top view of flex tray ice maker

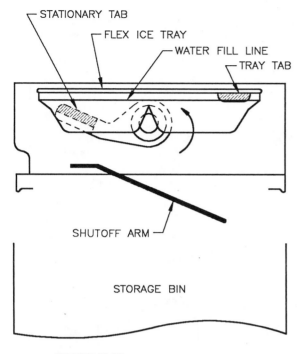

STATIONARY TAB

FLEX ICE TRAY

WATER FILL LINE

TRAY TAB

SHUTOFF ARM

STORAGE BIN

FIGURE 40-23

age bin. Notice that the time necessary to complete one cycle is about 13.3 minutes. When the turning tray approaches the upright position again, a cam operated switch energizes a water solenoid valve for a period of about 13 seconds and refills the tray with approximately 8 oz of water, figure 40-24.

FIGURE 40-24 Water solenoid switch

Note that replacement motors for this type of ice maker operate at a higher rate of speed than the original motors supplied with the unit. The new motors cause an ejection cycle to occur every 90 minutes instead of 120 minutes. For this reason, the water fill switch should be changed with the motor to prevent short cycling of the fill cycle.

The principle of operation of the flex tray ice maker is different than that of the compact type. The flex tray ice maker is incorporated in the same circuit with the defrost timer, figure 40-25. Since this type of ice maker does not need a mold heater or separate thermostat, it contains only three electrical components:

1. A timer motor, which operates both the defrost timer and the ice maker, figure 40-26. This motor contains a two-stage output gear. One gear operates the time cycle for the defrost heater, and the other gear operates the ice maker. The defrost cycle operates every 9.6 hours of timer motor running time, and the ice maker operates every 2 hours of timer motor running time. When the appliance is in the freeze cycle, the timer motor can operate only when both the cabinet thermostat and the defrost heater thermostat are closed. During the defrost cycle, however, the timer motor will continue to operate regardless of the condition of either thermostat.
2. Defrost timer switch, figure 40-27.
3. Water fill switch, figure 40-24.

Another operating difference is that the flex tray ice maker is more dependent on mechanical control than electrical control. The shutoff arm located at the bottom of the ice maker senses the level of ice in the storage bin. When this arm is in the down position, it permits a spring-loaded pin to move forward and lock a gear in place, figure 40-28. This locked gear permits the timing motor to turn the tray through one revolution and eject the ice into the storage bin. The pin is mechanically reset at the end of each ejection cycle.

When the shutoff arm is held up, the pin will not be released and the tray will not enter into an ejection cycle. The operation of the ice maker can be manually stopped by lifting the arm above its normal turnoff position.

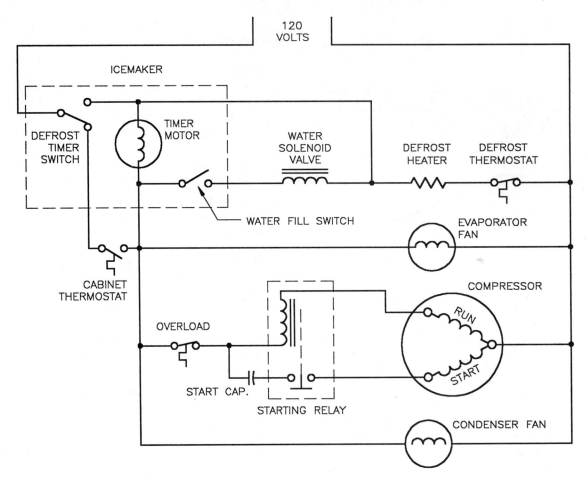

120 VOLTS

ICEMAKER

DEFROST TIMER SWITCH

TIMER MOTOR

WATER SOLENOID VALVE

DEFROST HEATER

DEFROST THERMOSTAT

WATER FILL SWITCH

EVAPORATOR FAN

CABINET THERMOSTAT

COMPRESSOR

RUN

START

OVERLOAD

START CAP.

STARTING RELAY

CONDENSER FAN

FIGURE 40-25

FIGURE 40-26 Timer motor

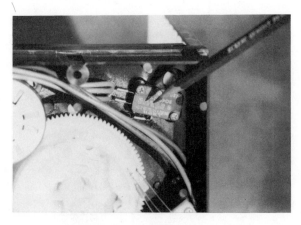

FIGURE 40-27 Defrost timer switch

SECTION 7 ICE MAKERS

FIGURE 40-28 Locking pin

CIRCUIT OPERATION

In the first stage of operation, the circuit is shown during the freeze cycle, figure 40-29. The cabinet thermostat and defrost heater thermostat are both closed. At this point, several circuit paths exist. One circuit is completed through the timer motor, defrost heater, and defrost thermostat. A circuit is completed through the evaporator fan, condenser fan, and the run winding of the compressor. If the cabinet thermostat should open, as shown in figure 40-30, the timer motor also stops.

After 9.6 hours of timer motor operation, the defrost timer switch changes position and completes

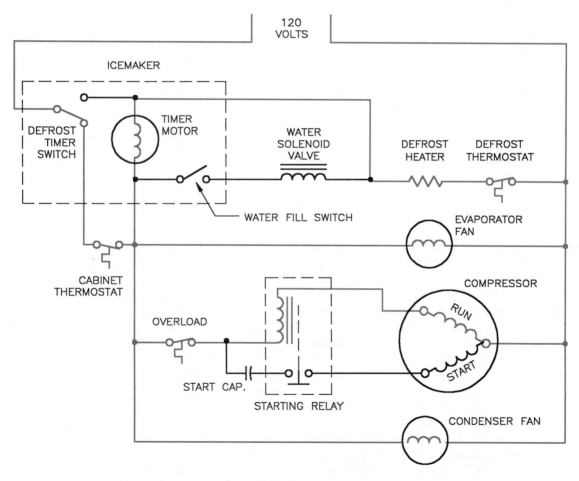

FIGURE 40-29

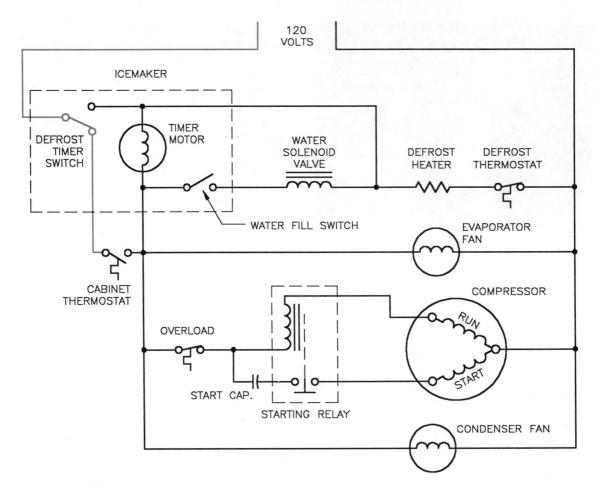

120 VOLTS

ICEMAKER

DEFROST TIMER SWITCH

TIMER MOTOR

WATER SOLENOID VALVE

DEFROST HEATER

DEFROST THERMOSTAT

WATER FILL SWITCH

EVAPORATOR FAN

CABINET THERMOSTAT

OVERLOAD

COMPRESSOR

RUN

START

START CAP.

STARTING RELAY

CONDENSER FAN

FIGURE 40-30

the circuits shown in figure 40-31. The defrost heater is now connected directly to the power line, which permits it to warm the evaporator and melt accumulations of frost. A current path also exists through the timer motor to the evaporator fan, compressor run winding, and condenser fan. It is this circuit path that permits the timer motor to continue operation if the defrost thermostat should open its contacts. The timer motor must continue to run, or the defrost cycle cannot be completed. Note that the winding of the timer motor has a much higher impedance than the run winding of the compressor. This permits almost all the voltage to be dropped across the timer motor and very little to be dropped across the evaporator fan, compressor run winding, and condenser fan. At this time, the timer motor will

operate, but the other motors will not. The defrost cycle lasts for approximately 21.5 minutes.

At the end of the defrost cycle, the defrost timer switch changes back to its normal position. Since the defrost thermostat opens its contacts at approximately 70°F and does not reclose them until the evaporator reaches about 2°F, it is normal for these contacts to be open at the end of the defrost cycle, figure 40-32. During this time, the timer motor is turned off.

Under normal conditions, the ice maker activates after 2 hours of timer motor running time. Figure 40-33 shows the condition of the circuit near the end of the ejection cycle. The cam-operated water fill switch has closed and now completes a circuit through the water solenoid valve, defrost heater, and

SECTION 7 ICE MAKERS

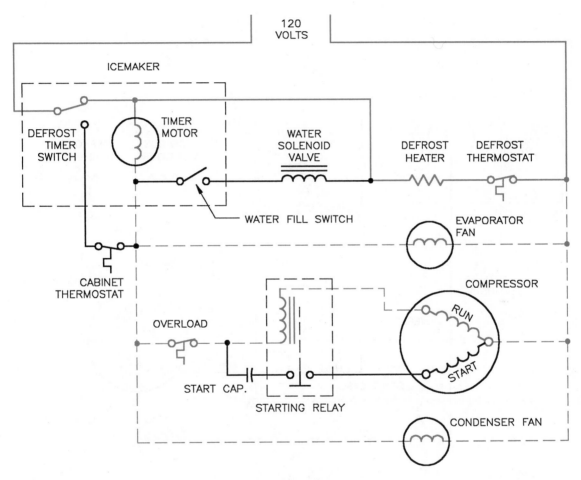

FIGURE 40-31

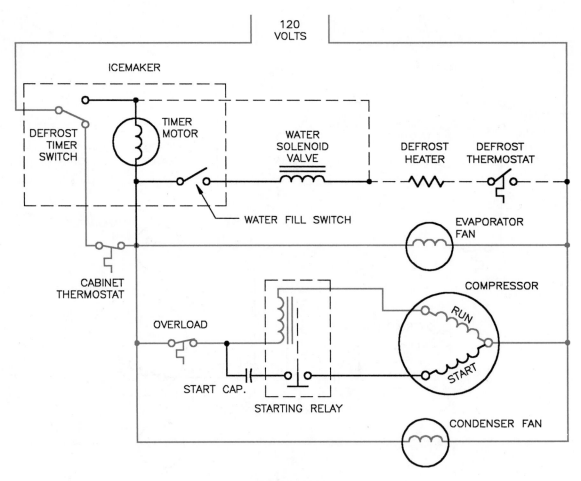

FIGURE 40-32

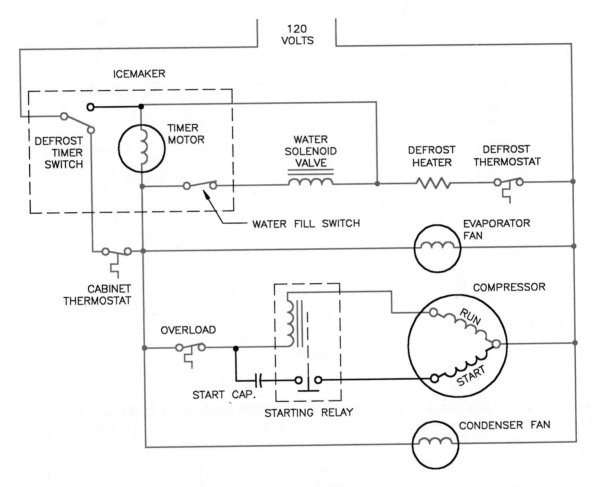

FIGURE 40-33

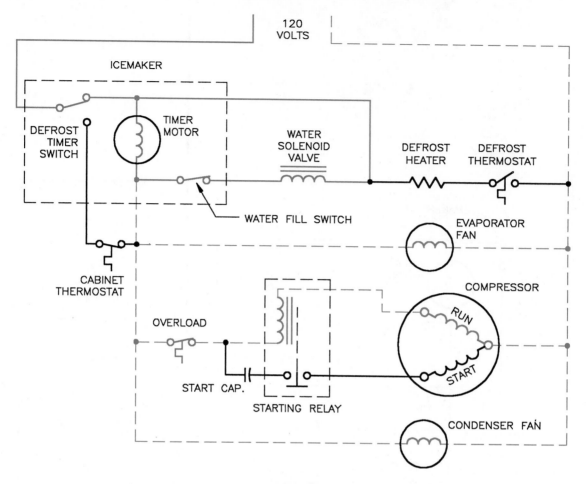

FIGURE 40-34

defrost thermostat. Note that if the cabinet thermostat should open during the water fill cycle, the cycle will be interrupted until the cabinet thermostat again closes.

Since the timer motor is used to operate both the ice maker and defrost cycle, it is possible for the ice ejection cycle to occur during the defrost cycle, figure 40-34. Notice that a parallel current path exists through both the timer motor and water solenoid valve. If the defrost thermostat should open while the water valve switch is closed, the same current path is provided through the evaporator fan, compressor run winding, and condenser fan for both the timer motor and water solenoid valve. This permits the tray to dump the ice and refill with water.

PROBLEMS AND PRECAUTIONS WITH THE FLEX TRAY ICE MAKER

When servicing this ice maker, there are several conditions the serviceman should be aware of:

1. Before disconnecting the ice maker, turn the cabinet thermostat to the **OFF** position or unplug the appliance. This is done to prevent arcing at the terminal connector block.
2. If the ice maker should jam, it will prevent the defrost timer from operating.
3. The flex tray type of ice maker permits no adjustment of the water fill switch.

4. One of the most common problems is that the surface of the ice tray becomes rough due to mineral deposits in the water. This causes the ice cubes to stick in the tray and not be ejected. When ice cubes are not ejected, the tray becomes too full during the fill cycle and causes a slab of ice to form. This slab can cause damage to the pin, gears, and/or timer motor. The ice tray should be replaced at the first sign of slabbing. A set of drive gears and pin are shown in figure 40-35.

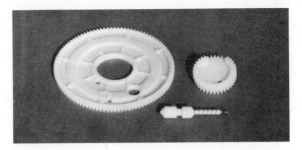

FIGURE 40-35 Drive gear and pins for the flex tray ice maker

REVIEW QUESTIONS

1. Ice makers are divided into what two major categories?

2. What is the advantage of continually recirculating the water during the ice making process?

3. What component controls the amount of water flow into the original compact ice maker?

4. Does the flex tray ice maker require a mold heater to thaw the ice cubes before they can be dumped into the storage bin?

5. In the original compact ice maker, what method is used to sense when the water has been frozen?

6. In the original compact ice maker, what controls the start of the ejection and refill cycle?

7. In the flex tray ice maker, what controls the start of the ejection and refill cycle?

8. How can the original compact ice maker be manually turned off?

9. How many revolutions will the ejector blades of the original compact ice maker normally make during the ejection cycle?

10. What is the function of the holding switch in the original compact ice maker circuit?

11. Concerning the flex tray ice maker, what two separate tasks are performed by the timer motor?

12. What is the function of the spring loaded pin in the flex tray ice maker?

13. Concerning the flex tray ice maker, is it possible for the timer motor to operate during the defrost cycle?

14. Concerning the new type compact ice maker, what method is used to change the contacts labeled A, B, C, and D in the schematic diagram shown in figure 40-18?

15. Can the gear of the new type compact ice maker be rotated to manually advance the operation of the ice maker?

16. How many revolutions do the ejector blades of the new type compact ice maker make during the ejection cycle?

UNIT 41
Commercial Ice Makers

Commercial ice makers are designed to produce large quantities of ice and are generally found in restaurants, cafeterias, motels, and hotels. Some ice makers produce cubes and others produce flaked ice. The first commercial type ice maker to be discussed is manufactured by SCOTSMAN Company. The basic components of this unit are shown in figure 41-1. Notice that this unit can be equipped with either an air-cooled or a water-cooled condenser. The water-cooled unit operates much more quietly.

This unit produces ice by cascading water over a metal plate used as the evaporator, figure 41-2. A water pump provides continuous circulation of water when the compressor is operating. A water distributor, located at the top of the plate, provides an even flow of water over the entire surface of the plate. Excess water is caught by a trough at the bottom of the plate and is returned to a sump where it is recirculated by the pump. Continuous circulation of water produces a clearer ice because pure water freezes faster than impure water. This is not to say that water can be purified by circulating it. The purification is a result of the freezing process. Basically, the water freezes before the impurities and minerals have a chance to freeze. The water, minus the impurities and minerals, is frozen to the evaporator plate and the impurities and minerals are returned to the sump.

After the ice has formed, the harvest cycle begins. At the beginning of this cycle, a hot-gas solenoid valve opens and permits high pressure hot gas to be diverted to the evaporator plate. This high pressure gas warms the plate, and thaws the surface of the ice which is in contact with it. The combination of the warm plate and the cascading water loosens the ice cubes so that they drop away from the plate and fall into the storage bin below. During the harvest cycle, a water solenoid valve opens and permits fresh water to flow into the sump. This not only refills the sump, but also flushes impurities out the overflow drain.

The basic electrical schematic diagram for this machine is shown in figure 41-3. There are two manual switches in this circuit. One is the master switch, and can be used to disconnect power to the entire control circuit. The second is connected in series with the compressor contactor relay coil. This switch can be used to turn off the compressor separately. Two safety switches, the high temperature cutout and the high pressure cutout, are connected in series with the master switch. If either of these switches opens, they will disconnect power to the control circuit.

This circuit also contains two thermostats, the bin thermostat, and the cube size thermostat. The sensor for the bin thermostat is located in the ice storage bin and senses the level of ice. When the ice reaches a high enough level, it touches the sensor and causes the contact to open. This stops the operation of the ice maker at the end of the harvest cycle. The sensor for the cube-size thermostat is mounted on the evaporator plate. When the evaporator plate reaches a low enough temperature, the thermostat

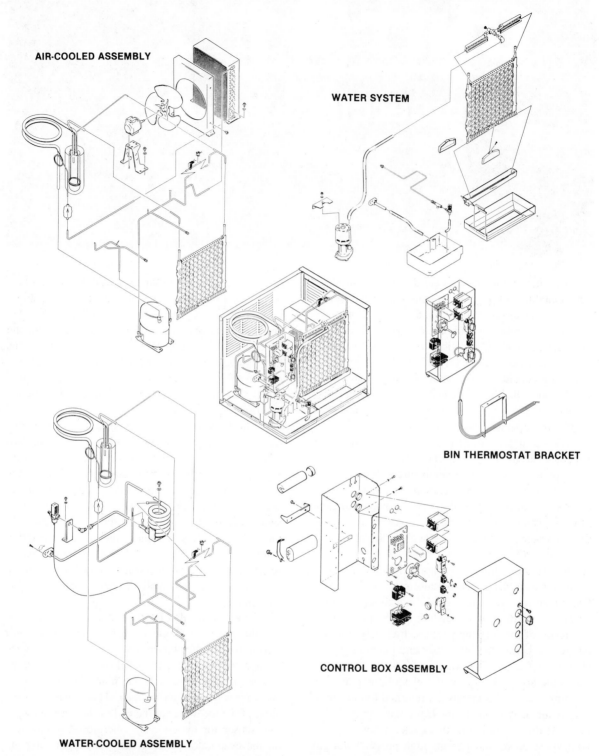

AIR-COOLED ASSEMBLY

WATER SYSTEM

BIN THERMOSTAT BRACKET

CONTROL BOX ASSEMBLY

WATER-COOLED ASSEMBLY

FIGURE 41-1 Scotsman cube-type ice maker (Courtesy Scotsman Ice Systems)

SECTION 7   ICE MAKERS

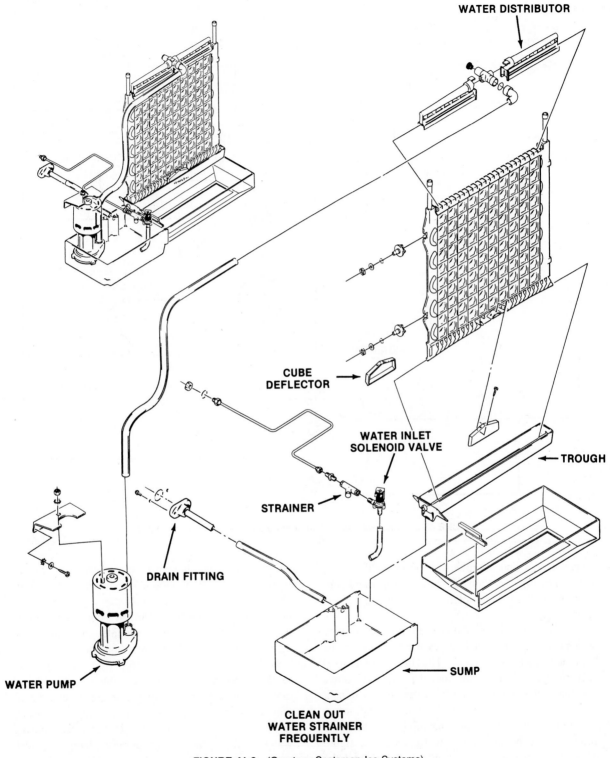

WATER DISTRIBUTOR

CUBE
DEFLECTOR

WATER INLET
SOLENOID VALVE

STRAINER

DRAIN FITTING

TROUGH

WATER PUMP

SUMP

CLEAN OUT
WATER STRAINER
FREQUENTLY

FIGURE 41-2 (Courtesy Scotsman Ice Systems)

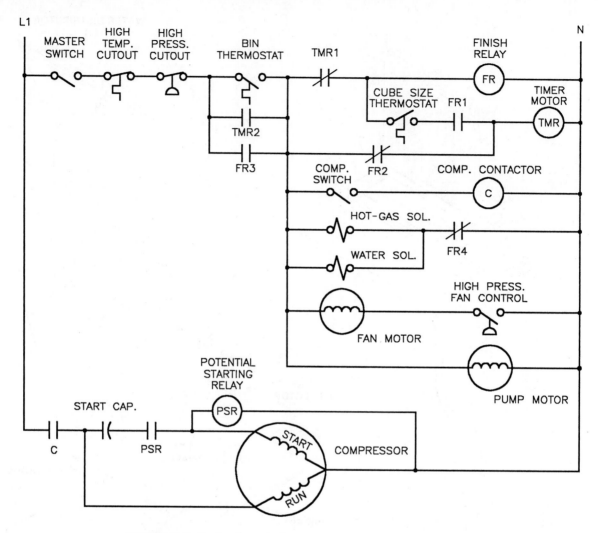

FIGURE 41-3 Basic schematic diagram of a commercial ice cube maker

contact closes and completes a circuit to the timer motor. The timer contains a set of cam-operated contacts and is used to complete the freeze and harvest cycle.

This circuit also contains a fan motor controlled by a pressure switch that senses the pressure on the high side of the compressor. This fan motor is used only on units with an air-cooled condenser. Since this fan motor is controlled by a pressure switch, it may cycle on and off during the unit's operation.

For a better understanding of this circuit, it is shown at the beginning of the freeze cycle, figure 41-4. It is assumed that the master switch and the

compressor switch have been closed, and that the bin thermostat contact is closed. A circuit is now completed to the finish relay, FR, causing all FR contacts to change position. The FR3 contact serves as a holding contact to permit completion of the cycle in the event the bin thermostat contact should open. The FR1 contact has closed to permit a current path to the timer motor when the cube size thermostat closes. The compressor contactor coil is energized, which closes its contact and connects the compressor to the line. The pump motor is energized causing water to flow over the evaporator plate. It is also assumed that the high-pressure fan control switch is

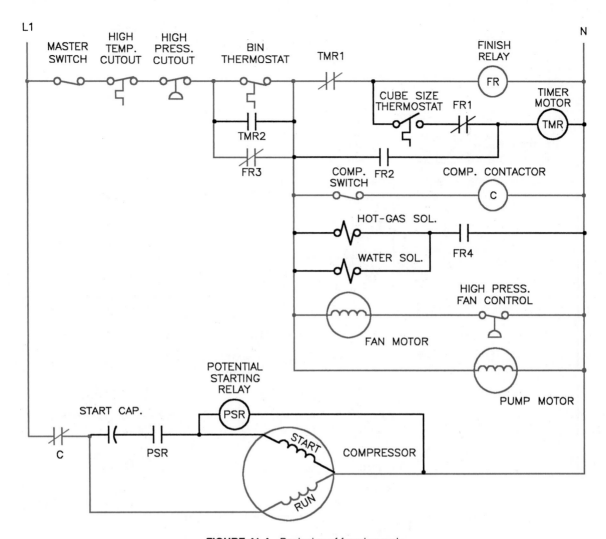

FIGURE 41-4 Beginning of freezing cycle

closed, permitting the condenser fan motor to operate. Notice, however, that the FR4 contact has opened to prevent the hot gas solenoid and the water solenoid from operating.

After the circuit has operated in this condition for some period of time, the evaporator plate becomes cold enough to permit the cube size thermostat to close as shown in figure 41-5. This completes a circuit to the timer motor. The timer is used to complete the cycle in the event the bin thermostat should open.

After the timer has operated for some length of time, the timer contacts will change position as

shown in figure 41-6, starting the harvest cycle. The TMR2 contact closes to maintain a current path around the bin thermostat contact, and the TMR1 contact opens and deenergizes coil FR. When coil FR deenergizes, all of its contacts return to their normal position. The FR2 contact recloses and maintains a current path to the timer motor, permitting it to complete the cycle. When the FR4 contact recloses, the hot-gas solenoid and water solenoid valves open. As hot gas is circulated through the evaporator plate, it warms and permits the cube size thermostat contact to reopen. The circuit will continue to operate in this condition until the timer com-

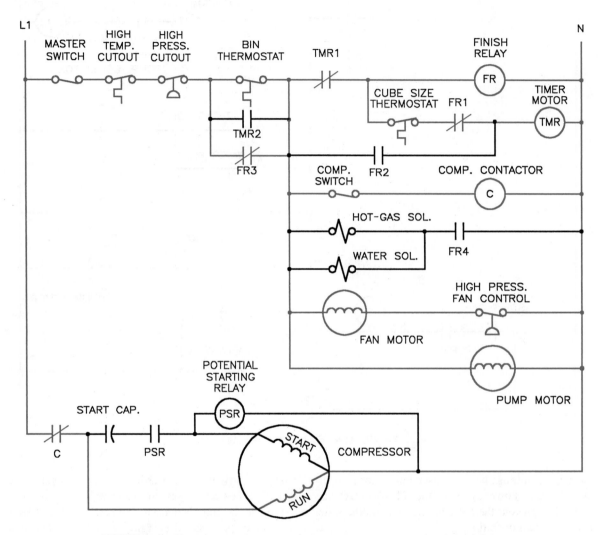

FIGURE 41-5 The cube size thermostat closes to complete a circuit to the timer motor.

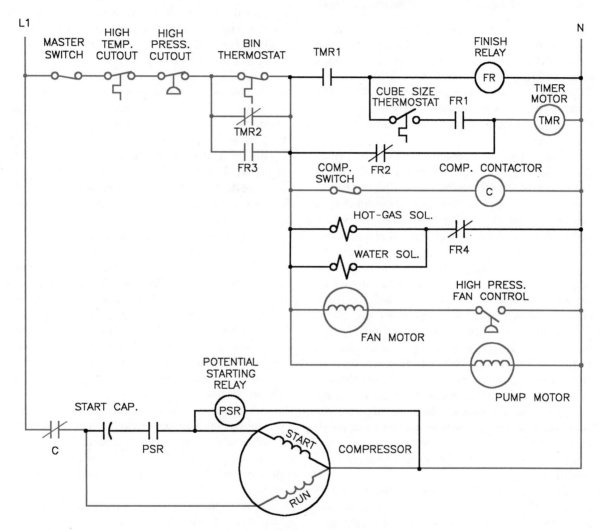

FIGURE 41-6 Harvest cycle

pletes the cycle and resets both TMR contacts. At this point, the freeze cycle will begin again if the bin thermostat is still closed.

FLAKER-TYPE ICE MAKERS

Flaker-type ice makers produce ice continuously as opposed to harvesting ice cubes at certain intervals. Flaked ice has a soft, flakey texture and is often preferred by restaurants. A basic diagram of a flaker-type ice maker is shown in figure 41-7. The water supply from the building enters the water reservoir. A float valve maintains a constant water level in the reservoir.

Water from the reservoir enters the bottom of the freezer assembly. The freezer assembly is the evaporator of the refrigeration unit. The freezer assembly is basically a hollow tube surrounded by a cylindrical container. Refrigerant is used to cool the hollow tube. A stainless steel auger is placed inside the hollow tube. The motor drive assembly turns the auger. As water enters the bottom of the freezer assembly, it is frozen into ice and carried upward by the auger. When the ice reaches the top, the flared end of the auger presses excess water out of the ice before it is extruded or flaked out through the ice spout. A nylobraid tube carries the ice to the ice storage bin. When enough ice accumulates, it touches the sensor bulb of the bin thermostat. The bin thermostat contact then opens and the compressor is disconnected from the line, but the auger drive motor continues to operate for approximately one to two minutes. This permits the auger to clear the ice out of the freezer unit before it stops operating.

The basic schematic diagram for the flaker-type ice maker is shown in figure 41-8. Notice that the auger drive motor contains two separate centrifugal switches, one normally closed and the other normally open. The normally closed switch connects the motor start winding to the line when the motor is started. The normally open centrifugal switch controls the coil of the compressor contactor. The compressor contactor can be energized only when the auger drive motor operates within a certain speed range. If ice becomes compacted in the freezer unit,

it will cause the auger drive motor to slow down. If the speed of the auger drive motor is reduced below a certain point, the centrifugal switch connected in series with the compressor contactor coil will open. If this should happen, the compressor turns off, but the auger delay pressure control switch permits the auger to continue operating for approximately one and a half minutes. If the ice is cleared sufficiently in that length of time, the auger drive motor speed will increase and permit the centrifugal switch to reclose and start the compressor.

The auger delay pressure control switch is a single-pole double-throw pressure switch connected in the low side of the refrigeration system. When the system is turned off, and the pressures have equalized in the system, the low-side pressure is high enough to hold the switch in the position shown in figure 41-8. When the compressor starts, the low-side pressure begins to decrease. When it has decreased to 20 PSIG (pounds per square inch gauge), the contacts change position. They will remain in the changed position until the low-side pressure increases to 32 PSIG.

The bin thermostat senses the level of ice in the storage bin and normally controls the operation of the ice maker. A low water-pressure switch is connected to the water supply line. If the water pressure drops below 5 PSIG, the switch contacts will open. They will reclose when the water pressure reaches 20 PSIG. The low head-pressure switch can interrupt operation of the compressor if the head pressure should become too low.

A master switch disconnects power to the entire control circuit. The spout switch can also disconnect power to the entire circuit in case the ice becomes compacted in the nylobraid tube and spout. If the spout switch becomes tripped, it must be manually reset.

This unit utilizes two condenser fan motors. One motor is mounted at the bottom of the condenser and the other is mounted at the top. The bottom fan motor is connected in parallel with the compressor and will operate any time the compressor is in operation. The top fan motor is controlled by a pressure switch that senses the high side of the refrigeration system. If the pressure becomes high enough, the switch contact will close and start the top condenser fan motor.

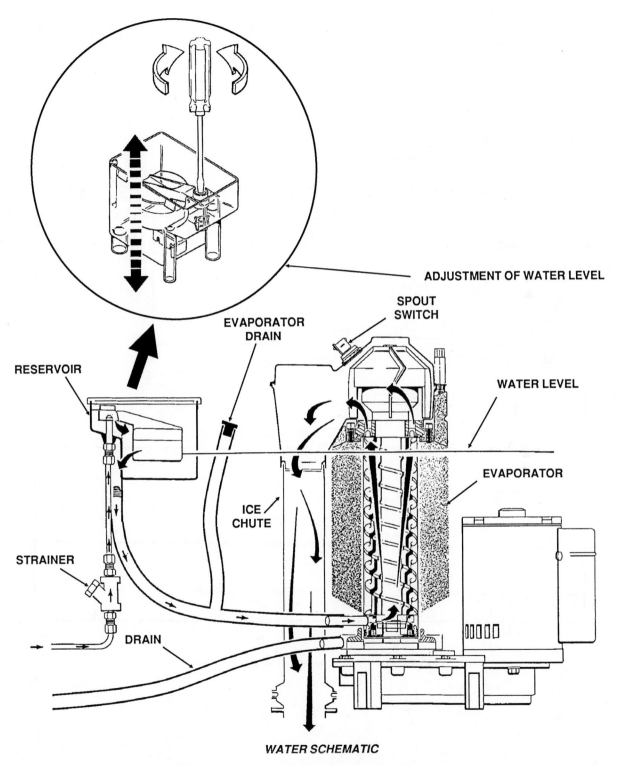

ADJUSTMENT OF WATER LEVEL

SPOUT
SWITCH

WATER LEVEL

EVAPORATOR
DRAIN

RESERVOIR

EVAPORATOR

ICE
CHUTE

STRAINER

DRAIN

WATER SCHEMATIC

FIGURE 41-7 Water schematic (Courtesy Scotsman Ice Systems)

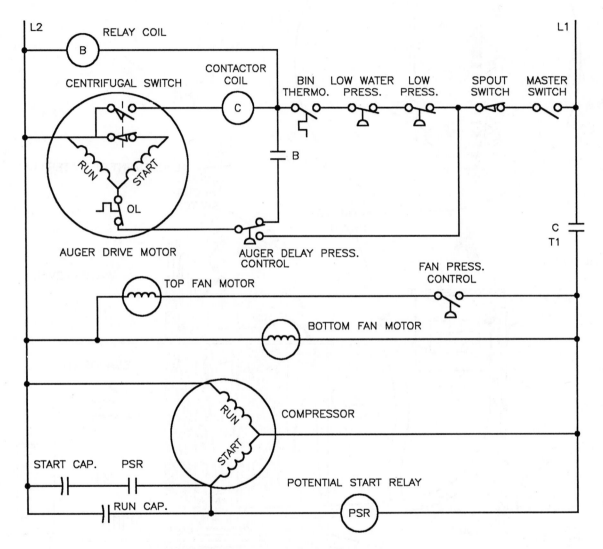

FIGURE 41-8 Basic schematic of flaker-type icemaker

OPERATION OF THE CIRCUIT

The circuit in figure 41-9 shows the initial start up of the ice maker. It is assumed that the master switch and the bin thermostat switch are closed. A circuit is first completed through coil B. This closes contact B and completes a circuit through the auger delay pressure control switch to the auger drive motor. The normally closed centrifugal switch contact completes a circuit to the start winding and permits the auger drive motor to start.

The normal running mode of the circuit is shown in figure 41-10. In this phase of operation, it is assumed that the auger drive motor is operating at the proper speed and the centrifugal switch connected in series with the compressor contactor has closed and permitted the compressor to start. The suction pressure has dropped low enough to permit the auger delay pressure control switch contacts to change position. The bottom condenser fan motor is in operation and the top fan motor may or may not be operating depending on the pressure on the high side of the refrigeration system.

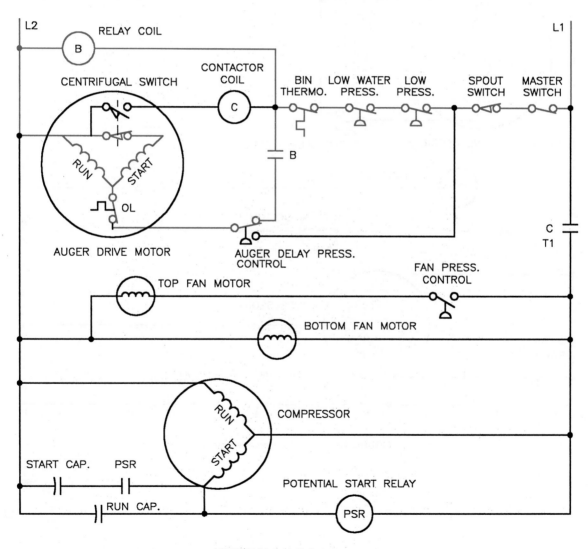

FIGURE 41-9 Initial start sequence

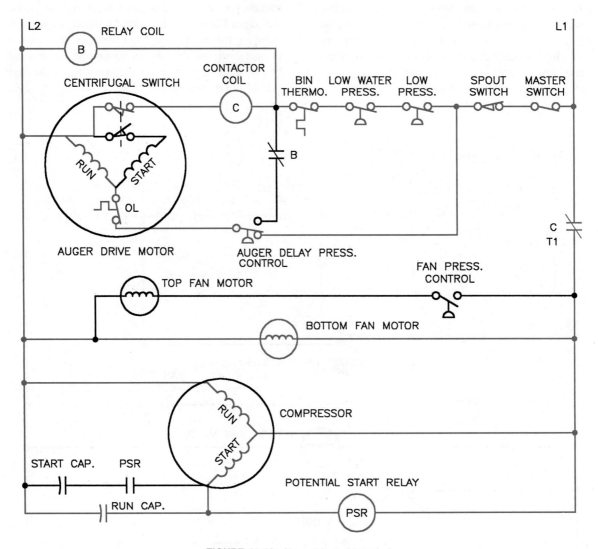

FIGURE 41-10 Normal icemaking mode

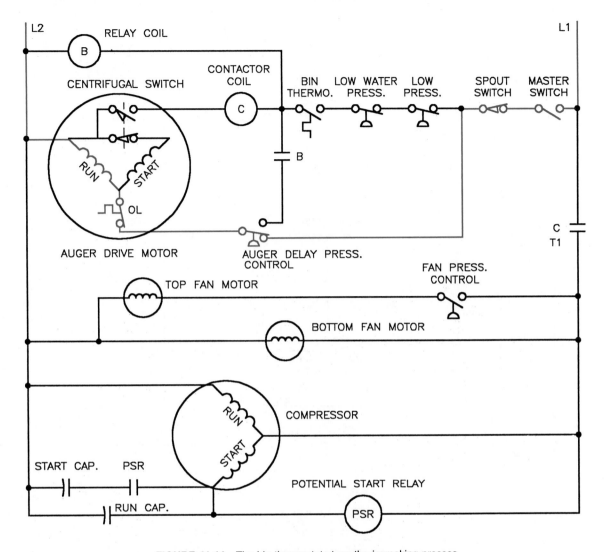

FIGURE 41-11 The bin thermostat stops the icemaking process.

In the schematic shown in figure 41-11, it is assumed that the bin thermostat has been satisfied and has opened its contact. This opens the circuit to coils B and C and stops the operation of the compressor. The auger drive motor will continue to operate until the pressure on the low side of the refrigeration system increases enough to reset the auger delay pressure switch. The circuit will then be back in its original, deenergized position.

The circuit in figure 41-12 shows the operation of the circuit when the auger drive motor slows down enough to cause the centrifugal switch in se- ries with the compressor contactor to open. If this occurs, the compressor will be disconnected from the line and stop operating. A current path is maintained through the auger drive motor and auger delay pressure switch. Notice also that a current path is maintained through relay coil B. If the auger drive motor speed does not increase sufficiently before the auger delay switch changes position, the closed B contact will provide a current path through the reset auger delay switch and permit the auger drive motor to continue operation.

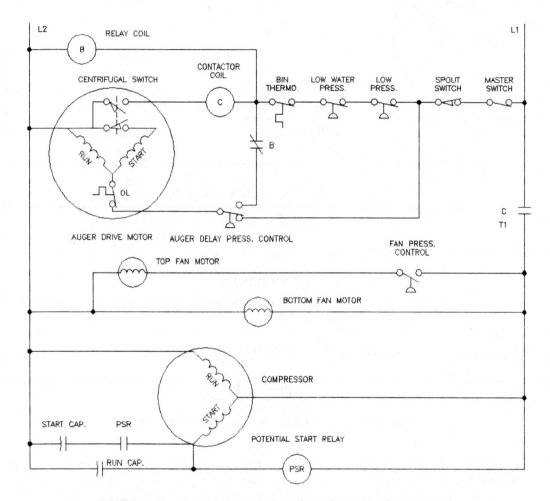

FIGURE 41-12 Auger becomes overloaded and disconnects the compressor.

REVIEW QUESTIONS

1. Concerning the Scotsman cube-type ice maker, what device is used to cause the water to flow evenly over the surface of the evaporator plate?

2. Concerning the Scotsman cube-type ice maker, what method is used to thaw the surface of the ice in contact with the evaporator plate during the harvest cycle?

3. Concerning the Scotsman cube-type ice maker, what are the two safety switches used to disconnect power from the control circuit?

4. What device is used to sense the level of ice cubes in the storage bin of the Scotsman cube-type ice maker?

5. What electrical component starts the operation of the timer motor in the Scotsman cube-type ice maker?

6. Concerning the Scotsman flaker-type ice maker, what device is used to carry the ice to the top of the evaporator tube?

7. How is excess water pressed out of the ice before it is ejected into the storage bin of the flaker-type ice maker?

8. Explain the operation of the auger delay switch used in the flaker-type ice maker.

9. Concerning the flaker-type ice maker, why is it desirable to have the auger drive motor continue to operate for some period of time after the compressor has stopped operation?

10. What controls the operation of the bottom condenser fan motor in the flaker-type ice maker?

11. What electrical component is used to stop the operation of the compressor if the auger should become overloaded?

12. Concerning the Scotsman flaker-type ice maker, which safety switch must be manually reset if it trips?

SECTION 8

Solid-State Devices

UNIT 42
Semi-Conductor Materials

Many of the air-conditioning controls are operated by solid-state devices as well as magnetic and mechanical devices. If a service technician is to install and troubleshoot control systems, he or she must have an understanding of electronic devices as well as relays.

Solid-state devices, such as diodes and transistors, are often referred to as *semi-conductors*. The word semi-conductor refers to the type of material solid-state devices are made of. To understand how solid-state devices operate, one must study the atomic structure of conductors, insulators, and semi-conductors.

CONDUCTORS

Conductors are materials that provide an easy path for electron flow. Conductors are generally made from materials that have large, heavy atoms. This is why most conductors are metals. The best electrical conductors are silver, copper, and aluminum. Conductors are materials that have only one or two valence electrons in their atom, figure 42-1. An atom that has only one valence electron makes the best electrical conductor because the electron is loosely held in orbit and is easily given up for current flow.

INSULATORS

Insulators are generally made from lightweight materials that have small atoms. The atoms of an insulating material will have their outer orbits filled or almost filled with valence electrons. This means an insulator will have seven or eight valence electrons, figure 42-2.

Since an insulator has its outer orbit filled or almost filled with valence electrons, they are tightly held in orbit and not easily given up for current flow.

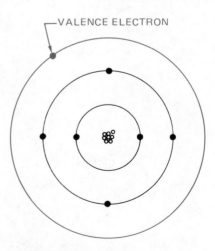

FIGURE 42-1   Atom of a conductor

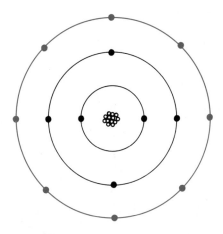

FIGURE 42-2 Atom on an insulator

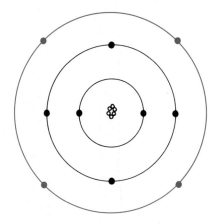

FIGURE 42-3 Atom of a semi-conductor

SEMI-CONDUCTORS

Semi-conductors, as the name implies, are materials that are neither good conductors nor good insulators. Semi-conductors are made from materials that have four valence electrons in their outer orbit, figure 42-3.

The most common semi-conductor materials used in the electronics field are germanium and silicon. Of these two materials, silicon is used more often because of its ability to withstand heat. When semi-conductor materials are refined into a pure form, the molecules arrange themselves into a crystal structure that has a definite pattern, figure 42-4. A pattern such as this is known as a *lattice structure*.

A pure semi-conductor material such as silicon has no special properties and will do little more than make a poor conductive material. If a semi-conductor material is to become useful for the production of solid-state components, it must be mixed with an impurity. When the semi-conductor material is mixed with an impurity that has only three

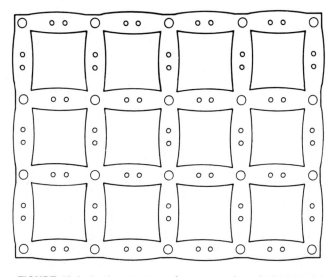

FIGURE 42-4 Lattice structure of a pure semi-conductor material

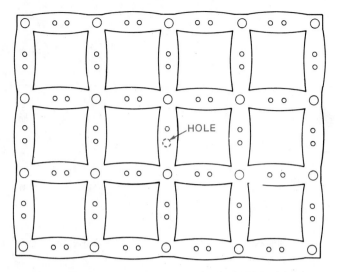

FIGURE 42-5 Lattice structure of a P-type material

valence electrons, such as idium or gallium, the lattice structure also becomes different, figure 42-5. When a material that has only three valence electrons is mixed with a pure semi-conductor, a hole is left in the material when the lattice structure is formed. This hole is caused by the lack of an electron where one should be. Since the material now has a lack of electrons, it is no longer electrically neutral. Electrons are negative particles. Since a hole is in a place where an electron should be, the hole has a positive charge. This semi-conductor material now has a net positive charge, and is therefore known as a P-type material.

When a semi-conductor material is mixed with an impurity that has five valence electrons, such as arsenic or antimony, the lattice structure will have an excess of electrons, figure 42-6. Since electrons are negative particles, and there are more electrons in the material than there should be, the material has a net negative charge. This material is referred to as an N-type material because of its negative charge.

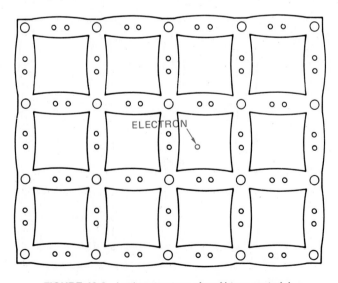

FIGURE 42-6 Lattice structure of an N-type material

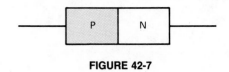

FIGURE 42-7

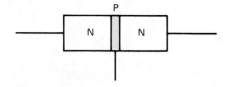

FIGURE 42-8 Transistor

All solid-state devices are made from a combination of P- and N-type materials. The type of device formed is determined by how the P- and N-type materials are connected or joined together. The number of layers of material and the thickness of various layers play an important part in determining what type of device will be formed. For instance, the diode is often called a PN junction because it is made by joining together a piece of P-type and a piece of N-type material, figure 42-7.

The transistor, on the other hand, is made by joining three layers of semi-conductor material, figure 42-8. Regardless of the type of solid-state device being used, it is made by the joining together of P- and N-type materials.

REVIEW QUESTIONS

1. How many valence electrons are contained in a material used as a conductor?

2. How many valence electrons are contained in a material used as an insulator?

3. What are the two most common materials used to produce semi-conductor devices?

4. What is a lattice structure?

5. How is a P-type material made?

6. How is an N-type material made?

7. What type of semi-conductor material can withstand the greatest amount of heat?

8. All solid-state components are formed from combinations of P- and N-type materials. What factors determine what kind of components will be formed?

UNIT 43

The PN Junction

As stated previously, solid-state devices are made by combining P- and N-type materials together. The device produced is determined by the number of layers of material used, the thickness of the layers of material, and the manner in which the layers are joined together. Hundreds of different electronic devices have been produced since the invention of solid-state components.

It is not within the scope of this text to cover even a small portion of these devices. The devices to be covered by this text have been chosen because of their frequent use in the air-conditioning industry as opposed to communications or computers. These devices are presented from a straightforward, practical viewpoint, and mathematical explanation is used only when necessary.

The PN junction is often referred to as the *diode*. The diode is the simplest of all electronic devices. It is made by joining together a piece of P-type material and a piece of N-type material. Refer to figure 43-1. The schematic symbol for a diode is shown in figure 43-2. The diode operates like an electric check valve in that it will permit current to flow through it in only one direction. If the diode is to conduct current, it must be forward biased. The diode is forward biased only when a positive voltage is connected to the anode and a negative voltage is connected to the cathode. If the diode is reverse biased, the negative voltage connected to the anode and the positive voltage connected to the cathode, it will act like an open switch and no current will flow through the device.

One thing the service technician should be aware of when working with solid-state circuits is that the explanation of the circuit is often given assuming conventional current flow as opposed to electron flow. *The conventional current flow theory assumes that current flows from positive to negative as opposed to the electron flow theory that states that current flows from negative to positive.* Although it has been known for many years that current flows from negative to positive, many of the electronic circuit explanations assume a positive to negative current flow. There are several reasons for this. For one, ground is generally negative and considered to be 0 volts in an electronic cir-

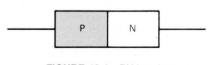

FIGURE 43-1 PN junction

FIGURE 43-2 Schematic symbol of a diode

cuit. Any voltage above or greater than ground is positive. Most people find it is easier to think of something flowing downhill or from some point above to some point below. Another reason is that all the arrows in an electronic schematic are pointed in the direction of conventional current flow. The diode shown in figure 43-2 is forward biased only when a positive voltage is applied to the anode and a negative voltage is applied to the cathode. If the conventional current flow theory is used, current will flow in the direction the arrow is pointing. If the electron theory of current flow is used, current must flow against the arrow.

A common example of the use of the conventional current flow theory is the electrical systems of automobiles. Most automobiles use a negative ground system, which means the negative terminal of the battery is grounded. The positive terminal of the battery is considered to be the "HOT" terminal, and it is generally assumed that current flows from the "HOT" to ground. This explanation is offered in an effort to avoid confusion when troubleshooting electronic circuits.

TESTING THE DIODE

The diode can be tested with an ohmmeter. When the leads of an ohmmeter are connected to a diode, the diode should show continuity in only one direction. For example, assume that when the leads of an ohmmeter are connected to a diode, it shows continuity. If the leads are reversed, the ohmmeter should indicate an open circuit. If the diode shows continuity in both directions, it is shorted. If the ohmmeter indicates no continuity in either direction, the diode is open. To test the diode, follow this two-step procedure:

1. Connect the ohmmeter leads to the diode. Notice if the meter indicates continuity through the diode or not, figure 43-3.
2. Reverse the diode connection to the ohmmeter, figure 43-4. Notice if the meter indicates continuity through the diode or not. The ohmmeter should indicate continuity through the diode in

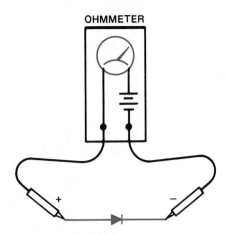

FIGURE 43-3 Testing a diode

only one direction. NOTE: If continuity is not indicated in either direction, the diode is open. If continuity is indicated in both directions, the diode is shorted.

RECTIFIERS

Diodes can be used to perform many jobs, but their most common use in industry is to construct a rectifier. A rectifier is a device that changes or converts AC voltage into DC voltage. The simplest type of rectifier is known as the half-wave rectifier.

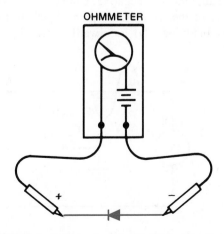

FIGURE 43-4 A diode connected in the reverse direction

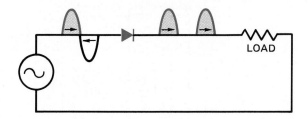

FIGURE 43-5 Half-wave rectifier

Refer to the circuit shown in figure 43-5. The half-wave rectifier can be constructed with only one diode, and gets its name from the fact that it will rectify only half of the AC waveform applied to it. When the voltage applied to the anode is positive, the diode is forward biased and current can flow through the diode, load resistor, and back to the power supply. When the voltage applied to the anode becomes negative, the diode is reverse biased and no current will flow. Since the diode permits current to flow through the load in only one direction, the current is DC.

Diodes can be connected to produce full-wave rectification, which means both halves of the AC waveform will be made to flow in the same direction. One type of full-wave rectifier is known as the bridge rectifier and is shown in figure 43-6. Notice the bridge rectifier requires 4 diodes for construction.

To understand the operation of the bridge rectifier, assume that point X of the AC source is positive and point Y is negative. Current will flow to point A of the rectifier. At point A, diode D4 is reverse biased and D1 is forward biased. The current will flow through diode D1 to point B of the rectifier. At point B, diode D2 is reverse biased, so the current must flow through the load resistor to ground. The current returns through ground to point D of the rectifier. At point D, both diodes D4 and D3 are forward biased, but current will not flow from positive to positive. Therefore, the current will flow through diode D3 to point C of the bridge, and then to point Y of the AC source, which is negative at this time. Since current flowed through the load resistor during this half cycle, a voltage is developed across the resistor.

Now assume that point Y of the AC source is positive and X is negative. Current will flow from point Y to point C of the rectifier. At point C, diode D3 is reverse biased and diode D2 is forward biased. The current will flow through diode D2 to point B of the rectifier. At point B, diode D1 is reverse biased, so the current must flow through the load resistor to ground. The current flows from ground to point D of the bridge. At point D, both diodes D3 and D4 are forward biased. As before, current will not flow from positive to positive, so the current will flow through diode D4 to point A of the bridge and then to point X, which is now negative. Since current flowed through the load resistor during this half cycle, a voltage is developed across the load resistor. Notice that the current flow was in the same direction through the resistor during both half cycles.

Most of industry operates on three-phase power instead of single-phase. Six diodes can be connected to form a three-phase bridge rectifier, which will change three-phase AC voltage into DC voltage. Refer to the circuit shown in figure 43-7.

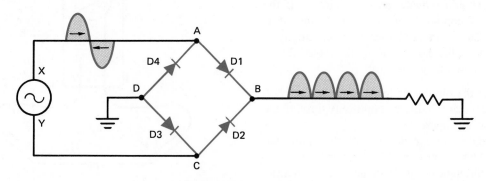

FIGURE 43-6 Bridge rectifier

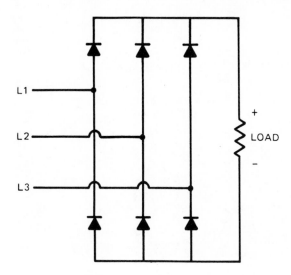

FIGURE 43-7   Three-phase bridge rectifier

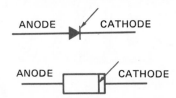

FIGURE 43-8   Lead identification of a plastic case diode

IDENTIFYING DIODE LEADS

When the diode is to be connected in a circuit, there must be some means of identifying the anode and the cathode. Diodes are made in different case styles so there are different methods of identifying the leads. Large stud-mounted diodes often have the diode symbol printed on the case to show proper lead identification. Small plastic case diodes often have a line or band around one end of the case, figure 43-8. This line or band represents the line in front of the arrow on the schematic symbol of the diode. An ohmmeter can always be used to determine the proper lead identification if the polarity of the ohmmeter leads is known. The positive lead of the ohmmeter must be connected to the anode to make the diode forward biased.

REVIEW QUESTIONS

1. What is the PN junction more commonly known as?

2. On a plastic case diode, how are the leads identified?

3. Explain how a diode operates.

4. Explain the difference between the conventional current flow theory and the electron flow theory.

5. Explain the difference between a half-wave rectifier and a full-wave rectifier.

6. Explain how to test a diode with an ohmmeter.

UNIT 44

The Light-Emitting Diode

The light-emitting diode, or LED as it is commonly referred to, is a special diode that emits light when current passes through it. The schematic symbol for the LED is shown in figure 44-1. Notice that the schematic symbol is the same as the junction diode, except that an arrow is pointing away from the diode. The arrow indicates that light is being given off by the device. Another device known as the photo diode is turned on by light. The schematic symbol for the photo diode is shown in figure 44-2. Notice the symbols are the same except that the arrow points toward the diode symbol for the photo diode. This shows that the device must receive light if it is to operate.

The light-emitting diode and the junction diode are similar in many respects. Both are rectifiers and both will permit current to flow in only one direction. The LED, however, has a higher voltage drop than the junction diode. For instance, a silicon junction diode must have about .7 volts to turn it on. The LED must have about 1.7 volts before it will be turned on. This higher voltage drop makes the LED difficult to test with an ohmmeter. An ohmmeter must produce enough voltage to turn the LED on, and many ohmmeters do not have this

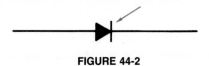

FIGURE 44-2

ability. The best method for testing the LED is to construct a circuit and see if it will operate.

When used in a circuit, the LED is generally operated at about 20 milliamps (ma) or less. For example, if an LED is to be connected to a 12-volt DC circuit, a current-limiting resistor must be connected in series with the LED, figure 44-3. The resistor needed is

$$R = E/R$$
$$R = 12/.020$$
$$R = 600 \text{ ohms}$$

The nearest standard 5% resistor value is 620 ohms.

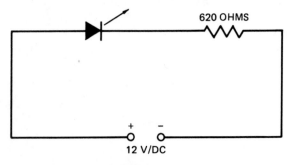

FIGURE 44-3

FIGURE 44-1

ANODE ▷|◁ CATHODE

FIGURE 44-4

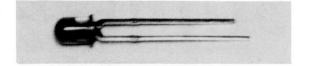

FIGURE 44-5 Light-emitting diode (LED)

Therefore, a 620-ohm resistor would be used to limit the current flow through the LED.

When connecting an LED into a circuit, it must be known which lead is the anode and which is the cathode. If the LED is held with the leads facing you, it can be seen that the case has a flat side close to one of the leads, figure 44-4. The flat side of the LED corresponds to the line in front of the arrow on the diode symbol.

The LED has become more and more popular in electronics since its invention. The LED is in-expensive and has no filament to burn out. LEDs are used as pilot lights on electronic equipment and as numerical displays. Many of the programmable thermostats use LEDs to indicate some process is in operation.

Another LED device that is becoming more and more popular in the air-conditioning and re-frigeration field is the solid-state relay. The LED is used in the opto-isolation circuit of the relay. This relay will be discussed later in the text. An LED is shown in figure 44-6.

REVIEW QUESTIONS

1. Will the LED rectify an AC voltage into a DC voltage?

2. What is the average voltage drop of an LED?

3. How can the anode and cathode of an LED be identified?

4. What is the average amount of current permitted to flow through an LED?

5. Can LEDs be tested with most ohmmeters?

UNIT 45

The Transistor

Transistors are made by joining three pieces of semi-conductor material together. There are two basic types of transistors, the NPN and the PNP, figure 45-1. The schematic symbols for these transistors are shown in figure 45-2. The difference in these transistors is the manner in which they are connected in a circuit. The NPN transistor must have positive connected to the collector and negative connected to the emitter. The PNP must have positive connected to the emitter and negative connected to the collector. Notice that the base must be connected to the same polarity as the collector to forward bias the transistor. Also notice that the arrows on the emitters point in the direction of conventional current flow.

TESTING THE TRANSISTOR

The transistor can be tested with an ohmmeter. If the polarity of the output of the ohmmeter leads is known, the transistor can be identified as either NPN or PNP. A transistor will appear to an ohmmeter to be two diodes joined together, figure 45-3. An NPN transistor will appear to an ohmmeter to be two diodes with their anodes connected

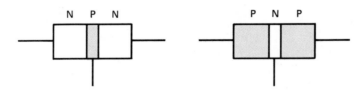

FIGURE 45-1   Two basic types of transistors

+ BASE   + COLLECTOR   NPN   – EMITTER

– BASE   – COLLECTOR   PNP   + EMITTER

FIGURE 45-2   Schematic symbols of transistors

FIGURE 45-3 Ohmmeter test for transistors

together. If the positive lead of the ohmmeter is connected to the base of the transistor, a diode junction should be seen between the base-collector and the base-emitter. If the negative lead of the ohmmeter is connected to the base of an NPN transistor, there should be no continuity between the base-collector and the base-emitter junction.

A PNP transistor will appear to be two diodes with their cathodes connected together. If the negative lead of the ohmmeter is connected to the base of the transistor, a diode junction should be seen between the base-collector and the base-emitter. If the positive ohmmeter lead is connected to the base, there should be no continuity between the base-collector or the base-emitter.

The following step-by-step procedure can be used to test a transistor.

1. Using a diode, determine which ohmmeter lead is positive and which is negative. The ohmmeter will indicate continuity through the diode only when the positive lead is connected to the anode and the negative lead is connected to the cathode, figure 45-4.

2. If the transistor is an NPN, connect the positive ohmmeter lead to the base and the negative lead to the collector. The ohmmeter should indicate continuity. The reading should be about the same as the reading obtained when the diode was tested, figure 45-5.

3. With the positive ohmmeter lead still connected to the base of the transistor, connect the negative lead to the emitter. The ohmmeter should again indicate a forward diode junction, figure 45-6. NOTE: If the ohmmeter does not indicate continuity between the base-collector or the base-emitter, the transistor is open.

4. Connect the negative ohmmeter lead to the base and the positive lead to the collector. The ohmmeter should indicate infinity or no continuity, figure 45-7.

5. With the negative ohmmeter lead connected to the base, reconnect the positive lead to the emitter. There should, again, be no indication of continuity, figure 45-8. NOTE: If a very high resistance is indicated by the ohmmeter, the transistor is "leaky" but it may still operate in

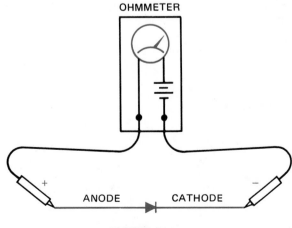

FIGURE 45-4

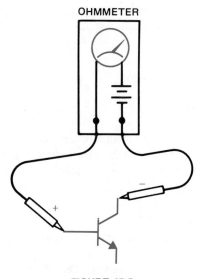

FIGURE 45-5

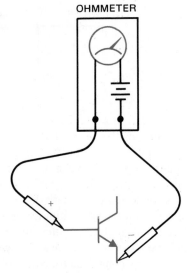

FIGURE 56-6

the circuit. If a very low resistance is seen, the transistor is shorted.

6. To test a PNP transistor, reverse the polarity of the ohmmeter leads and repeat the test. When the negative ohmmeter lead is connected to the base, a forward diode junction should be indicated when the positive lead is connected to the collector or emitter, figure 45-9.

7. If the positive ohmmeter lead is connected to

the base of a PNP transistor, no continuity should be indicated when the negative lead is connected to the collector or the emitter, figure 45-10.

TRANSISTOR OPERATION

The simplest way to describe the operation of a transistor is to say it operates like an electric valve.

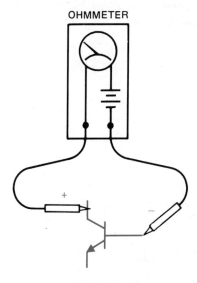

FIGURE 45-7

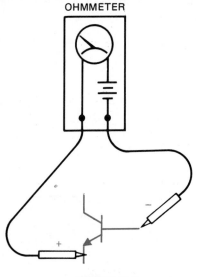

FIGURE 45-8

SECTION 8   SOLID-STATE DEVICES

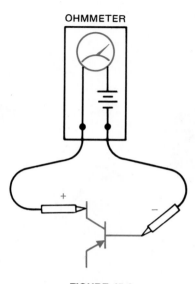

FIGURE 45-9

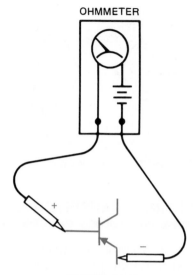

FIGURE 45-10

Current will not flow through the collector-emitter until current flows through the base-emitter. The amount of base-emitter current, however, is small when compared to the collector-emitter current, figure 45-11. For example, assume that when one milliamp (ma) of current flows through the base-emitter junction, 100 ma of current flows through the collector-emitter junction. If this transistor is a linear device, an increase or decrease of base current will cause a similar increase or decrease of collector current. For instance, if the base current is increased to 2 ma, the collector current would increase to 200 ma. If the base current is decreased to .5 ma, the collector current would decrease to 50 ma. Notice that a small change in the amount of base current can cause a large change in the amount of collector current. This permits a small amount of signal current to operate a larger device, such as the coil of a control relay.

One of the most common applications of the transistor is that of a switch. When used in this manner, the transistor operates like a digital device instead of an analog device. The term digital refers to a device that has only two states such as on or off. An analog device can be adjusted to different states. An example of this control can be seen in a simple switch connection. A common wall switch is a digital device. It can be used to either turn a

light on or off. If the simple toggle switch is replaced with a dimmer control, the light can be turned on, off, or adjusted to any position between. The dimmer is an example of analog control.

If no current flows through the base of the transistor, the transistor acts like an open switch and no current will flow through the collector-emitter junction. If enough base current is applied to the transistor to turn it completely on, it acts like a closed switch and permits current to flow through the collector-emitter junction. This is the same action produced by the closing contacts of a relay or motor starter, but a relay or motor starter cannot turn on and off several thousand times a second and a transistor can.

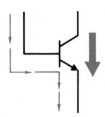

FIGURE 45-11 A small base current controls a large collector current.

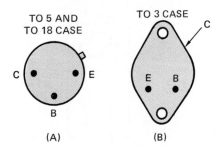

FIGURE 45-12 Lead identification of transistors

IDENTIFYING TRANSISTOR LEADS

Some case styles of transistors permit the leads to be quickly identified. The TO 5 and TO 18 cases, and the TO 3 case are in this category. The leads of the TO 5 or TO 18 case transistors can be identified by holding the case of the transistor with the leads facing you as shown in figure 45-12(A). The metal tab on the case of the transistor is closest to the emitter lead. The base and collector leads are positioned as shown in figure 45-12(A).

The leads of a TO 3 case transistor can be

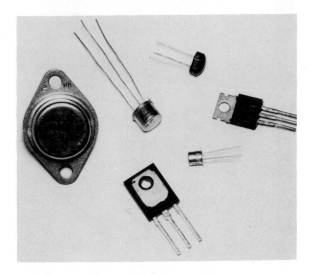

FIGURE 45-13 Transistors shown in different case styles

identified as shown in figure 45-12(B). With the transistor held with the leads facing you and down, the emitter is the left lead and the base is the right lead. The case of the transistor is the collector. Several case styles for the transistor are shown in figure 45-13.

REVIEW QUESTIONS

1. What are the two basic types of transistors?

2. Explain how to test an NPN transistor with an ohmmeter.

3. Explain how to test a PNP transistor with an ohmmeter.

4. What polarity must be connected to the collector, base, and emitter of an NPN to make it forward biased?

5. What polarity must be connected to the collector, base, and emitter of a PNP transistor to make it forward biased?

6. Explain the difference between an analog device and a digital device.

UNIT 46

The Unijunction Transistor

The unijunction transistor is a special transistor that has two bases and one emitter. The unijunction transistor (UJT) is a digital device because it has only two states, on or off, and is generally classified with a group of devices known as *thyristors*. Thyristors are turned completely on or completely off. Thyristors include devices such as the SCR, Triac, Diac, and the UJT.

The unijunction transistor is made by combining three layers of semi-conductor material as shown in figure 46-1. The schematic symbol with polarity connections and the base diagram is shown in figure 46-2.

UJT CHARACTERISTICS

The UJT has two paths for current flow. One path is from B2 to B1. The other path is through the emitter and base #1. In its normal state, there

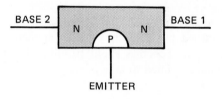

FIGURE 46-1 Unijunction transistor

is no current flow through either path until the voltage applied to the emitter reaches about 10 volts higher than the voltage applied to base #1. When the voltage applied to the emitter reaches about 10 volts more positive than the voltage applied to base #1, the UJT turns on and current flows through the B1–B2 path and from the emitter through base #1. Current will continue to flow through the UJT until the voltage applied to the emitter drops to a point that it is only about 3 volts higher than the voltage applied to B1. When the emitter voltage

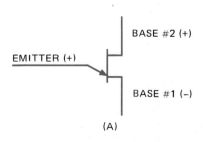

(A)

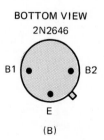

(B)

FIGURE 46-2 The schematic symbol for the unijunction transistor with polarity connections and base diagram

293

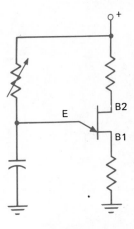

FIGURE 46-3

has been charged to about 10 volts, the UJT turns on and discharges the capacitor through the emitter and base #1. When the capacitor has been discharged to about 3 volts, the UJT turns off and permits the capacitor to begin charging again. By varying the resistance connected in series with the capacitor, the amount of time needed for charging the capacitor can be changed, thereby controlling the pulse rate of the UJT ($T = RC$).

The unijunction transistor can furnish a large output pulse, because the output pulse is produced by the discharging capacitor, figure 46-4. This large output pulse is generally used for triggering the gate of an SCR.

The pulse rate is determined by the amount of resistance and capacitance connected to the emitter of the UJT. However, the amount of capacitance that can be connected to the UJT is limited. For instance, most UJTs should not have a capacitor larger than 10 μf connected to them. If the capacitor is too large, the UJT will not be able to handle the current spike produced by the capacitor, and the UJT could be damaged.

drops to this point, the UJT will turn off and remain turned off until the voltage applied to the emitter again becomes about 10 volts higher than the voltage applied to B1.

CIRCUIT OPERATION

The unijunction transistor is generally connected into a circuit similar to the circuit shown in figure 46-3. The variable resistor controls the rate of charge time of the capacitor. When the capacitor

TESTING THE UJT

The unijunction transistor can be tested with an ohmmeter in a manner very similar to testing a common junction transistor. The UJT will appear

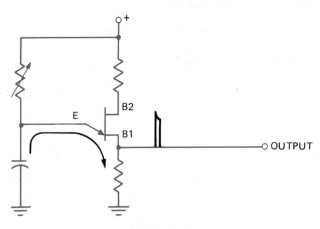

FIGURE 46-4

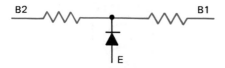

FIGURE 46-5 The UJT appears as two resistors connected to a diode when tested with an ohmmeter.

to the ohmmeter to be a connection of two resistors connected to a common junction diode. The common junction point of the two resistors will appear to be at the emitter of the UJT as shown in figure 46-5. When the positive lead of the ohmmeter is connected to the emitter, a diode junction should be seen from the emitter to base #2 and another diode connection from the emitter to base #1. If the negative lead of the ohmmeter is connected to the emitter of the UJT, no connection should be seen between the emitter and either base.

The following step-by-step procedure can be used to test a unijunction transistor.

1. Using a junction diode, determine which ohmmeter lead is positive and which is negative. The ohmmeter will indicate continuity when the positive lead is connected to the anode and the negative lead is connected to the cathode, figure 46-6.
2. Connect the positive ohmmeter lead to the emitter lead and the negative lead to base #1. The

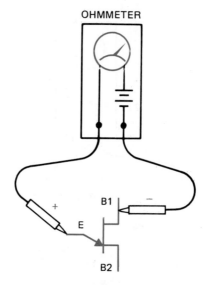

FIGURE 46-7

ohmmeter should indicate a forward diode junction, figure 46-7.
3. With the positive ohmmeter lead connected to the emitter, reconnect the negative lead to base #2. The ohmmeter should again indicate a forward diode junction, figure 46-8.
4. If the negative ohmmeter lead is connected to the emitter, no continuity should be indicated when the positive lead is connected to base #1 or base #2, figure 46-9.

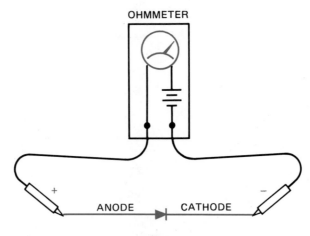

FIGURE 46-6

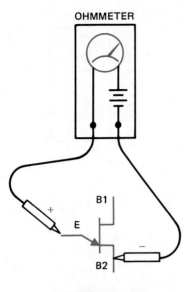

FIGURE 46-8

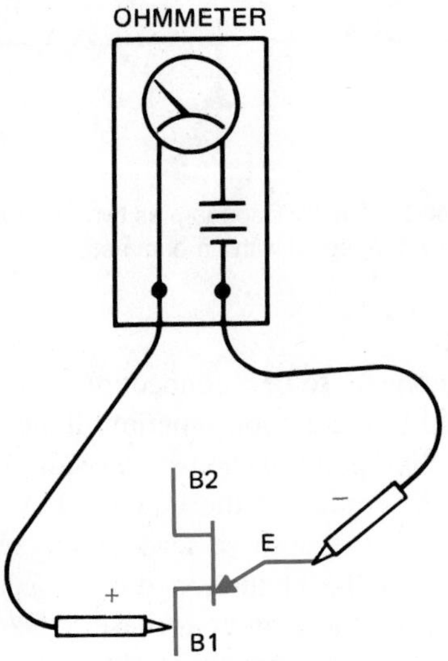

FIGURE 46-9

REVIEW QUESTIONS

1. What do the letters UJT stand for?

2. How many layers of semi-conductor material are used to construct a UJT?

3. Briefly explain the operation of the UJT.

4. Draw the schematic symbol for the UJT.

5. Briefly explain how to test a UJT with an ohmmeter.

UNIT 47

The Silicon-Controlled Rectifier

The silicon-controlled rectifier (SCR) is often referred to as the PNPN junction because it is made by joining 4 layers of semi-conductor material together, figure 47-1. The schematic symbol for the SCR is shown in figure 47-2. Notice that the symbol for the SCR is the same as the diode except that a gate lead has been added.

SCR CHARACTERISTICS

The SCR is a member of a family of devices known as thyristors. Thyristors are digital devices in that they have only two states, on or off. The SCR is used when it is necessary for an electronic device to control a large amount of power. Assume an SCR has been connected in a circuit as shown in figure 47-3. When the SCR is turned off, it will drop the full voltage of the circuit and 200 volts will appear across the anode and cathode. Although the SCR has a voltage drop of 200 volts, there is no current flow in the circuit. The SCR does not have to dissipate any power in this condition (200 volts × 0 amps = 0 watts). When the

pushbutton is pressed, the SCR will turn on. When the SCR turns on, it will have a voltage drop across its anode and cathode of about 1 volt. The load resistor limits the circuit current to 2 amps (200 volts/100 ohms = 2 amps). Since the SCR now has a voltage drop of 1 volt and 2 amps of current is flowing through it, it must now dissipate 2 watts of heat (1 volt × 2 amps = 2 watts). Notice that the SCR is dissipating only 2 watts of power, but is controlling 200 watts.

THE SCR IN A DC CIRCUIT

When an SCR is connected in a DC circuit as shown in figure 47-3, the gate will turn the SCR on but will not turn the SCR off. The gate must be connected to the same polarity as the anode if it is to turn the anode-cathode section of the SCR on. Once the gate has turned the SCR on, it will remain turned on until the current flowing through the anode-cathode drops to a low enough level to permit the device to turn off. The amount of current required to keep the SCR turned on is called

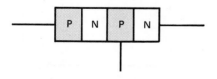

FIGURE 47-1 PNPN junction

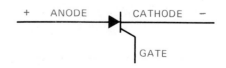

FIGURE 47-2 Schematic symbol of an SCR

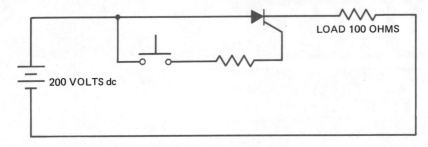

FIGURE 47-3 Gate turns SCR on.

the *holding current,* figure 47-4. Assume resistor R1 has been adjusted for its highest value and resistor R2 has been adjusted to its lowest or 0 value. When switch S1 is closed, no current will flow through the anode-cathode section of the SCR because resistor R1 prevents enough current flowing through the gate-cathode section of the SCR to trigger the device. If resistor R1 is slowly decreased in value, current flow through the gate-cathode will slowly increase. When the gate current reaches a certan level, assume 5 ma for this SCR, the SCR will fire or turn on. When the SCR fires, current will flow through the anode-cathode section and the voltage drop across the device becomes about 1 volt. Once the SCR has turned on, the gate has no more control over the device and could be disconnected from the anode without having any effect on the circuit. When the SCR fires, the anode-cathode becomes a short circuit for all practical purposes and current flow is limited by resistor R3. Now assume that resistor R2 is slowly increased in value. When the resistance of R2 is slowly increased, the current flow through the anode-cathode will slowly decrease. Assume that when the current flow through the anode-cathode drops

to 100 ma, the device suddenly turns off and the current flow drops to 0. This SCR requires 5 ma. of gate current to turn it on, and has a holding current value of 100 ma.

THE SCR IN AN AC CIRCUIT

The SCR is a rectifier. When it is connected in an AC circuit the output will be DC. The SCR operates in the same manner in an AC circuit as it does in a DC circuit. The difference in operation is caused by the AC waveform falling back to 0 at the end of each half cycle. When the AC waveform drops to 0 at the end of each half cycle, it will permit the SCR to turn off. This means the gate must re-trigger the SCR for each cycle it is to conduct. Refer to the circuit shown in figure 47-5.

Assume that the varible resistor connected to the gate has been adjusted to permit 5 ma of current to flow when the voltage applied to the anode reaches its peak value. When the SCR turns on, current will begin flowing through the load resistor when the AC waveform is at its positive peak. Current will continue to flow through the load until the

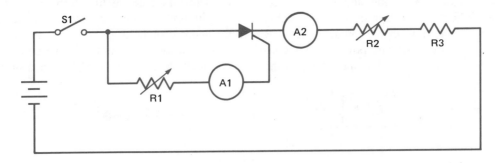

FIGURE 47-4 Operation of an SCR in a DC circuit

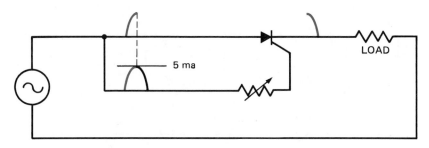

FIGURE 47-5   SCR fires when the AC waveform reaches peak value.

decreasing voltage of the sine wave causes the current to drop below the holding current level of 100 ma. When the current through the anode-cathode drops below 100 ma, the SCR truns off and all current flow stops. The SCR will remain turned off when the AC waveform goes into the negative half cycle because it is reverse biased and cannot be fired.

If the resistance connected in series with the gate is reduced, a current of 5 ma will be reached before the AC waveform reaches it peak value, figure 47-6. This causes the SCR to fire sooner in the cycle. Since the SCR fires sooner, current is permitted to flow through the load resistor for a longer period of time, which causes a higher average voltage drop across the load. If the resistance of the gate circuit is reduced again, as shown in figure 47-7, the 5 ma of gate current needed to fire the SCR will be reached sooner than before. This permits current to begin flowing through the load sooner than before, which permits a higher average voltage to be dropped across the load.

Notice that this circuit will permit the SCR to control only half of the positive waveform. The latest the SCR can be fired in the cycle is when the AC waveform is at 90° or peak. If a lamp were used as the load for this circuit, it would burn at half brightness when the SCR first turned on. This control would permit the lamp to be operated from half brightness to full brightness, but it could not be operated at a level less than half brightness.

PHASE SHIFTING THE SCR

If the SCR is to control all of the positive waveform, it must be phase shifted. As the term implies, phase shifting means to shift the phase of one thing in reference to another. In this instance, the voltage applied to the gate must be shifted out of phase with the voltage applied to the anode. There are several methods that can be used for phase shifting an SCR, but it is beyond the scope of this text to cover all of them. The basic principles are

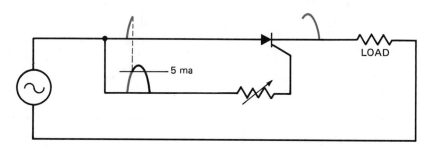

FIGURE 47-6   SCR fires before the AC waveform reaches peak value.

UNIT 47   THE SILICON-CONTROLLED RECTIFIER

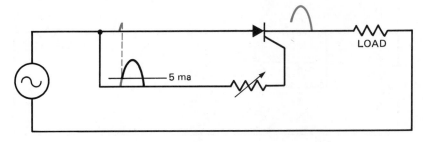

FIGURE 47-7 SCR fires sooner than before.

the same for all of the methods, however, so only one method will be covered.

If an SCR is to be phase shifted, the gate circuit must be unlocked or separated from the anode circuit. The circuit shown in figure 47-8 will accomplish this. A 24-volt center-tapped transformer has been used to isolate the gate circuit from the anode circuit. Diodes D1 and D2 are used to form a two-diode type of full wave rectifier to operate the unijunction transistor (UJT) circuit. Resistor R1 is used to determine the pulse rate of the UJT by controlling the charge time of capacitor C1. Resistor R2 is used to limit the current through the emitter of the UJT if resistor R1 is adjusted to 0 ohms. Resistor R3 limits current through the Base 1–Base 2 section when the UJT turns on. Resistor R4 permits a voltage spike or pulse to be produced across it when the UJT turns on and discharges ca-

pacitor C1. The pulse produced by the discharge of capacitor C1 is used to trigger the gate of the SCR.

Since the pulse of the UJT is used to provide a trigger for the gate of the SCR, the SCR can now be fired at any time regardless of the voltage applied to the anode. This means the SCR can now be fired as early or late during the positive half cycle as desired, because the gate pulse is now determined by the charge rate of capacitor C1. The voltage across the load can now be adjusted from 0 to the full applied voltage.

TESTING THE SCR

The SCR can be tested with an ohmmeter. To test the SCR, connect the positive output lead of

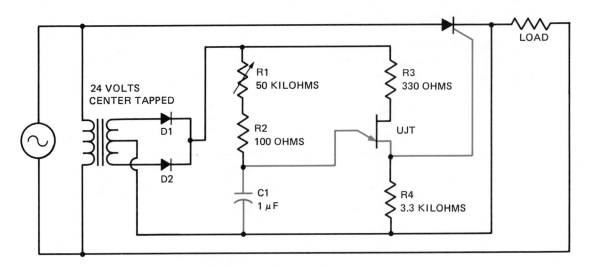

FIGURE 47-8 UJT phase shift for an SCR

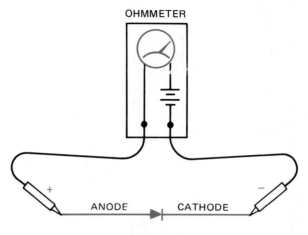

FIGURE 47-9

the ohmmeter to the anode and the negative lead to the cathode. The ohmmeter should indicate no continuity. Touch the gate of the SCR to the anode. The ohmmeter should indicate continuity through the SCR. When the gate lead is removed from the anode, conduction may stop or continue, depending on whether the ohmmeter is supplying enough current to keep the device above its holding current level or not. If the ohmmeter indicates continuity through the SCR before the gate is touched to the anode, the SCR is shorted. If the ohmmeter will not indicate continuity through the SCR after the gate has been touched to the anode, the SCR

is open. The following step-by-step procedure can be used for testing an SCR.

1. Using a junction diode, determine which ohmmeter lead is positive and which is negative. The ohmmeter will indicate continuity only when the positive lead is connected to the anode of the diode and the negative lead is connected to the cathode, figure 47-9.

2. Connect the positive ohmmeter lead to the anode of the SCR and the negative lead to the cathode. The ohmmeter should indicate no continuity, figure 47-10.

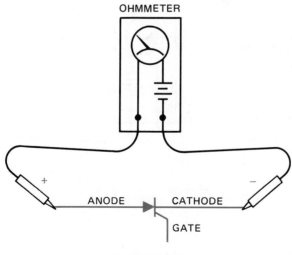

FIGURE 47-10

UNIT 47   THE SILICON-CONTROLLED RECTIFIER

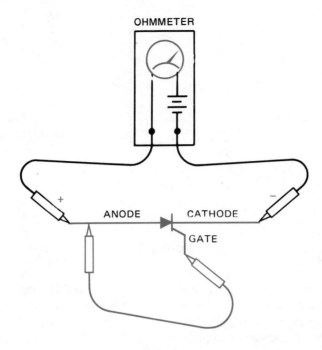

FIGURE 47-11

3. Using a jumper lead, connect the gate of the SCR to the anode. The ohmmeter should indicate a forward diode junction when the connection is made, figure 47-11. NOTE: If the jumper is removed, the SCR may continue to conduct or it may turn off. This will be determined by whether or not the ohmmeter can supply enough current to keep the SCR above its holding current level.

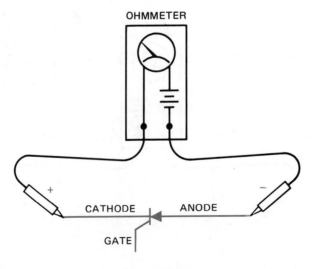

FIGURE 47-12

　　　　　　　　　　　　　　　　SECTION 8   SOLID-STATE DEVICES

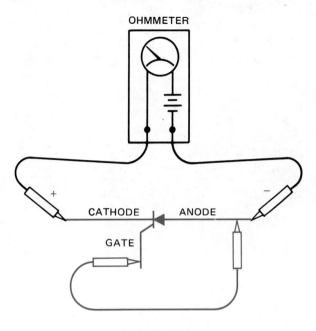

OHMMETER

CATHODE   ANODE

GATE

FIGURE 47-13

4. Reconnect the SCR so that the cathode is connected to the positive ohmmeter lead and the anode is connected to the negative lead. The ohmmeter should indicate no continuity, figure 47-12.

5. If a jumper lead is used to connect the gate to the anode, the ohmmeter should indicate no continuity, figure 47-13. NOTE: SCRs designed to switch large current (50 amperes or more) may indicate some leakage current with this test. This is normal for some devices.

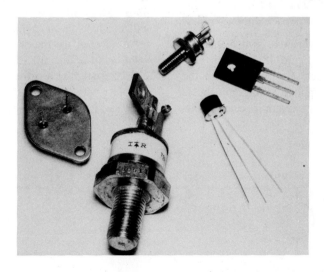

FIGURE 47-14   SCRs shown in different case styles

UNIT 47   THE SILICON-CONTROLLED RECTIFIER

UNIT 48

The Diac

The diac is a special-purpose bidirectional diode. The primary function of the diac is to phase shift a triac. The operation of the diac is very similar to that of a unijunction transistor, except the diac is a "bi" or two-directional device. The diac has the ability to operate in an AC circuit while the UJT is a DC device only.

There are two schematic symbols for the diac, figure 48-1. Either of these symbols is used in an electronic schematic to illustrate the use of a diac; therefore, you should become familiar with both.

DIAC CHARACTERISTICS

The diac is a voltage-sensitive switch that can operate on either polarity, figure 48-2. When voltage is applied to the diac, it will remain in the turned-off state until the applied voltage reaches a predetermined level. For this example, assume this

FIGURE 48-1 Schematic symbols for a diac

to be 15 volts. When the voltage reaches 15 volts, the diac will turn on or fire. When the diac fires, it displays a negative resistance, which means it will conduct at a lower voltage than the voltage that was applied to it, assume 5 volts. The diac will remain turned on until the applied voltage drops below its conduction level, which in this example is 5 volts. Refer to the waveform shown in figure 48-3. Since the diac is a bidirectional device, it will conduct on either half cycle of the AC applied to it. Refer to the waveform shown in figure 48-4. Notice that the diac has the same operating characteristic with either half cycle of AC. The simplest way to sum up the operation of the diac is to say it is a voltage-sensitive AC switch.

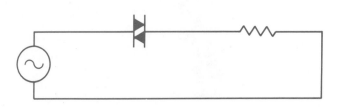

FIGURE 48-2 The diac can operate on either polarity

FIGURE 48-3 The diac operates until the applied voltage falls below its conduction level.

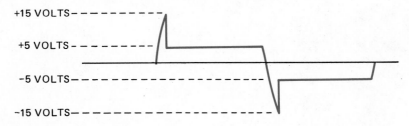

FIGURE 48-4 The diac will conduct on either half of the alternating current.

REVIEW QUESTIONS

1. Briefly explain how a diac operates.

2. Draw the two schematic symbols for the diac.

3. What is the major use of the diac in industry?

4. When a diac first turns on, does the voltage drop, remain at the same level, or increase to a higher level?

UNIT 49

The Triac

The triac is a PNPN junction connected in parallel with an NPNP junction. Figure 49-1 illustrates the semi-conductor arrangement of a triac. The triac operates similar to two SCRs in parallel, facing in opposite directions with their gate leads connected together, figure 49-2. The schematic symbol for the triac is shown in figure 49-3.

When an SCR is connected in an AC circuit, the output voltage will be DC. When a triac is connected in an AC circuit, the output voltage will be AC. Since the triac operates like two SCRs connected together and facing in opposite directions, it will conduct both the positive and negative half cycles of AC current.

When a triac is connected in an AC circuit as shown in figure 49-4, the gate must be connected to the same polarity as MT2. When the AC voltage applied to MT2 becomes positive, the SCR, which is forward biased, will conduct. When the voltage applied to MT2 becomes negative, the other SCR is forward biased and will conduct that half of the waveform. Since one of the SCRs is forward biased

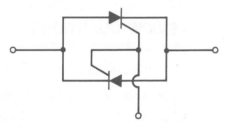

FIGURE 49-2 The triac operates similar to two SCRs with a common gate.

for each half cycle, the triac will conduct AC current as long as the gate lead is connected to MT2.

The triac, like the SCR, requires a certain amount of gate current to turn it on. Once the triac has been triggered by the gate, it will continue to conduct until the current flowing through MT2-MT1 drops below the holding current level.

THE TRIAC USED AS AN AC SWITCH

The triac is a member of the thyristor family and has only two states of operation, on or off.

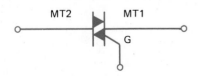

FIGURE 49-3 Schematic symbol of a triac

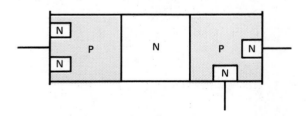

FIGURE 49-1 Semi-conductor arrangement of a triac

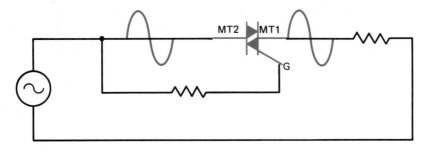

FIGURE 49-4 The triac will conduct both halves of the AC waveform.

When the triac is turned off it will drop the full applied voltage of the circuit at 0 amps of current flow. When the triac is turned on, it has a voltage drop of about 1 volt and circuit current must be limited by the load connected to the circuit. The triac has become very popular in industrial circuits as an AC switch. Since it is a thyristor, it has the ability to control a large amount of voltage and current. There are no contacts to wear out, it is sealed against dirt and moisture, and it can operate thousands of times per second. The triac is used as the output of many solid-state relays that will be covered later.

THE TRIAC USED FOR AC VOLTAGE CONTROL

The triac can be used to control an AC voltage, figure 49-5. If a variable resistor is connected in series with the gate, the point at which the gate current will reach a high enough level to fire the triac can be adjusted. The resistance can be adjusted to permit the triac to fire when the AC waveform reaches its peak value. This will cause half of the AC voltage to be dropped across the triac and half to be dropped across the load.

If the gate resistance is reduced, the amount of gate current needed to fire the triac will be obtained before the AC waveform reaches its peak value. This means that less voltage will be dropped across the triac and more voltage will be dropped across the load. This circuit permits the triac to control only one half of the AC waveform applied to it. If a lamp is used as the load, it can be con-

trolled from half brightness to full brightness. If an attempt is made to adjust the lamp to operate at less than half brightness, it will turn off.

PHASE SHIFTING THE TRIAC

The triac, like the SCR, must be phase shifted if complete voltage control is to be obtained. There are several methods that can be used to phase shift a triac, but only one will be covered in this unit. In this example, a diac will be used to phase shift the triac, figure 49-6. In this circuit, resistors R1 and R2 are connected in series with capacitor C1. Resistor R1 is a variable resistor and is used to control the charge time of capacitor C1. Resistor R2 is used to limit current if resistor R1 should be adjusted to 0 ohms. Assume the diac connected in series with the gate of the triac will turn on when capacitor C1 has been charged to 15 volts. When the diac turns on, capacitor C1 will discharge through the gate of the triac. This permits the triac to fire or turn on.

Once the triac has fired, there will be a voltage drop of about 1 volt across MT2 and MT1. The triac will remain turned on until the AC voltage drops to a low enough value to permit the triac to turn off. Since the phase shift circuit is connected in parallel with the triac, once the triac turns on capacitor C1 cannot begin charging again until the triac turns off at the end of the AC cycle. The diac, being a bidirectional device, will permit a positive or negative pulse to trigger the gate of the triac.

Notice that the pulse applied to the gate is controlled by the charging of capacitor C1 and not

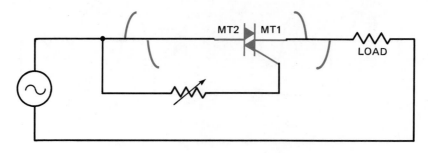

FIGURE 49-5 The triac controls half of the AC applied voltage.

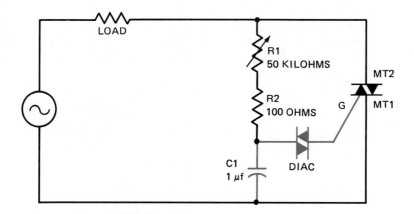

FIGURE 49-6 Phase-shift circuit for a triac

the amplitude of voltage. If the correct values are chosen, the triac can be fired at any point in the AC cycle applied to it. The triac can now control the AC voltage from 0 to the full voltage of the circuit. A common example of this type of triac circuit is the light dimmer control used in many homes.

TESTING THE TRIAC

The triac can be tested with an ohmmeter. To test the triac, connect the ohmmeter leads to MT2 and MT1. The ohmmeter should indicate no continuity. If the gate lead is touched to MT2, the triac should turn on and the ohmmeter will indicate continuity through the triac. When the gate lead is released from MT2, the triac may continue to conduct or turn off depending on whether the ohmmeter supplies enough current to keep the device above

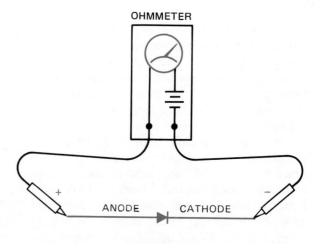

FIGURE 49-7

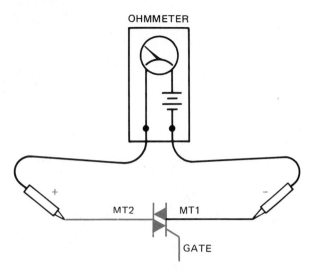

FIGURE 49-8

its holding current level. This tests one half of the triac. To test the other half of the triac, reverse the connection of the ohmmeter leads. The ohmmeter should again indicate no continuity. If the gate is touched again to MT2, the ohmmeter should indicate continuity through the device. The other half of the triac has been tested. The following step-by-step procedure can be used to test a triac.

1. Using a junction diode, determine which ohmmeter lead is positive and which is negative. The ohmmeter will indicate continuity only when the positive lead is connected to the anode and the negative lead is connected to the cathode, figure 49-7.
2. Connect the positive ohmmeter lead to MT2 and the negative lead to MT1. The ohmmeter should indicate no continuity through the triac, figure 49-8.
3. Using a jumper lead, connect the gate of the triac to MT2. The ohmmeter should indicate a forward diode junction, figure 49-9.
4. Reconnect the triac so that MT1 is connected to the positive ohmmeter lead and MT2 is connected to the negative lead. The ohmmeter should indicate no continuity through the triac, figure 49-10.

5. Using a jumper lead, again connect the gate to MT2. The ohmmeter should indicate a forward diode junction, figure 49-11.

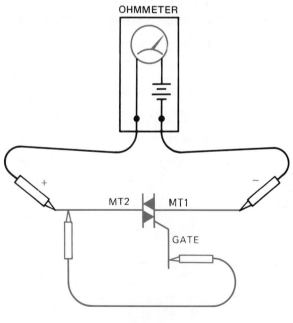

FIGURE 49-9

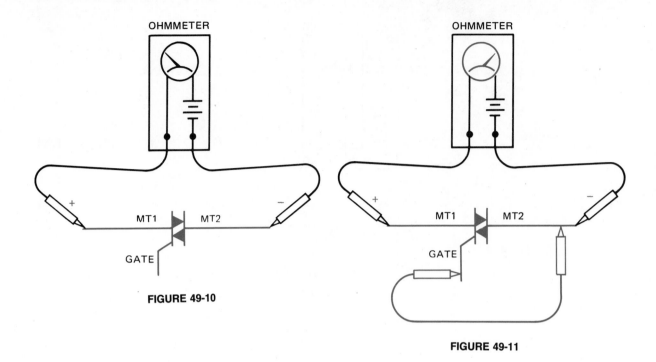

FIGURE 49-10

FIGURE 49-11

REVIEW QUESTIONS

1. Draw the schematic symbol of a triac.

2. When a triac is connected in an AC circuit, is the output AC or DC?

3. The triac is a member of what family of devices?

4. Briefly explain why a triac must be phase shifted.

5. What electronic component is frequently used to phase shift the triac?

6. When the triac is being tested with an ohmmeter, which other terminal should the gate be connected to if the ohmmeter is to indicate continuity?

UNIT 50

The Operational Amplifier

The operational amplifier has become another very common component found in industrial electronic circuits. The operational amplifier, or *op amp* as it is generally referred to, is used in hundreds of different applications. There are different types of op amps used, depending upon the type of circuit it is intended to operate in. Some op amps use bipolar transistors for the input and others use field effect transistors. The advantage of using field effect transistors is their extremely high input impedance, which can be several thousand megohms. The advantage of this extremely high input impedance is that it does not require a large amount of current to operate the amplifier. In fact, op amps, which use FET inputs, are generally considered as requiring no input current.

THE IDEAL AMPLIFIER

Before continuing the discussion of op amps, it should first be decided what an ideal amplifier is. First, the ideal amplifier should have an input impedance of infinity. If the amplifier had an input impedance of infinity, it would require no power drain on the signal source being amplified. Therefore, regardless of how weak the input signal source is, it would not be affected when connected to the amplifier. The ideal amplifier would have 0 output impedance. If the amplifier had 0 output impedance it could be connected to any load resistance

desired and not drop any voltage inside the amplifier. If it had no internal voltage drop, the amplifier would utilize 100% of its gain. Third, the amplifier would have unlimited gain. This would permit it to amplify any input signal as much as desired.

741 PARAMETERS

There is no such thing as the ideal or perfect amplifier of course, but the op amp can come close. One of the old reliable op amps, which is still used to a large extent, is the 741. The 741 will be used in this description as a typical operational amplifier. Please keep in mind that there are other op amps that have different characteristics of input and output impedance, but the basic theory of operation is the same for all of them.

The 741 op amp uses bipolar transistors for the input. The input impedance is about 2 megohms, and the output impedance is about 75 ohms. Its open loop or maximum gain is about 200,000. Actually, the 741 op amp has such a high gain that it is generally impractical to use and negative feedback, which will be discussed later, is used to reduce the gain. For instance, assume the amplifier has an output voltage of 15 volts. If the input signal voltage is greater than 1/200,000 of the output voltage or 75 microvolts (15/200,000 = .000075), the amplifier would be driven into saturation at which point it would not operate.

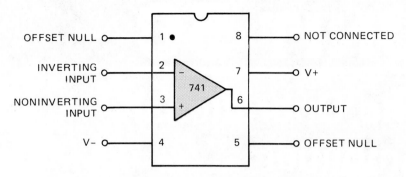

FIGURE 50-1 741 operational amplifier

741 PIN CONNECTION

The 741 operational amplifier is generally housed in an 8-pin in-line IC package, figure 50-1. Pins #1 and #5 are connected to the offset null. The offset null is used to produce 0 volts at the output. What happens is this: The op amp has two inputs called the inverting input and the noninverting input. These inputs are connected to a differential amplifier which amplifies the difference between the two voltages. If both of these inputs are connected to the same voltage, say by grounding both inputs, the output should be 0 volts. In actual practice, however, there are generally unbalanced conditions in the op amp that cause a voltage to be produced at the output. Since the op amp has a very high gain, a very slight imbalance of a few microvolts at the input can cause several millivolts at the output. The offset nulls are adjusted after the 741 is connected into a working circuit. Adjustment is made by connecting a 10K ohm potentiometer across pins #1 and #5, and connecting the wiper to the negative voltage, figure 50-2.

Pin #2 is the inverting input. If a signal is applied to this input, the output will be inverted. For instance, if a positive-going AC voltage is applied to the inverting input, the output will produce a negative-going voltage, figure 50-3.

Pin #3 is the noninverting input. When a signal voltage is applied to the noninverting input, the output voltage will be the same polarity. If a pos-

itive-going AC signal is applied to the noninverting input, the output voltage will be positive also, figure 50-4.

Pins #4 and #7 are the voltage input pins. Operational amplifiers are generally connected to above- and below-ground power supplies. These power supplies produce both a positive and negative voltage as compared to ground. There are some circuit connections that do not require an above- and below-ground power supply, but these are the exception instead of the rule. Pin #4 is connected to the negative- or below-ground voltage and pin #7 is connected to the positive- or above-ground voltage. The 741 will operate on voltages that range from about 4 volts to 16 volts. Generally, the operating voltage for the 741 is 12 to 15 volts plus and minus. The 741 has a maximum power output rating of about 500 milliwatts. Pin #6 is the output and pin #8 is not connected.

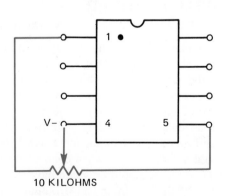

FIGURE 50-2 Offset null connection

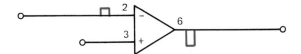

FIGURE 50-3 Inverted output

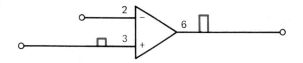

FIGURE 50-4 Negative feedback connection

NEGATIVE FEEDBACK

As stated previously, the open loop gain of the 741 operational amplifier is about 200,000. This amount of gain is not practical for most applications, so something must be done to reduce this gain to a reasonable level. One of the great advantages of the op amp is the ease with which the gain can be controlled, figure 50-5. The amount of gain is controlled by a negative-feedback loop. This is accomplished by feeding a portion of the output voltage back to the inverting input. Since the output voltage is always opposite in polarity to the inverting input voltage, the amount of output voltage fed back to the input tends to reduce the input voltage. Negative feedback has two effects on the operation of the amplifier. One effect is that it reduces the gain. The other is that it makes the amplifier more stable.

The gain of the amplifier is controlled by the ratio of resistors R2 and R1. If a noninverting amplifier is used, the gain is found by the formula (R2 + R1)/R1. If resistor R1 is 1K ohms and resistor R2 is 10K ohms, the gain of the amplifier would be 11 (11,000 ÷ 1,000 = 11).

If the op amp is connected as an inverting amplifier, however, the input signal will be out of phase with the feedback voltage of the output. This will cause a reduction of the input voltage applied to the amplifier and a reduction in gain. The formula (R2/R1) is used to compute the gain of an inverting amplifier. If resistor R1 is 1K ohms and resistor R2 is 10K ohms, the gain of the inverting amplifier would be 10 (10,000 ÷ 1,000 = 10).

There are some practical limits, however. As a general rule, the 741 operational amplifier is not operated above a gain of about 100. If more gain is desired, it is generally obtained by using more than one amplifier, figure 50-6.

As shown in figure 50-6, the output of one amplifier is fed into the input of another amplifier. The reason for not operating the 741 at high gain is that at high gains it tends to become unstable. Another general rule for operating the 741 op amp is the total feedback resistance (R1 + R2) is usually kept more than 1000 ohms and less than 100,000 ohms. These general rules apply to the 741 operational amplifier and may not apply to other operational amplifiers.

BASIC CIRCUIT CONNECTIONS

Op amps are generally used in three basic ways. This is not to say that op amps are used in only three circuits, but that there are three basic circuits that are used to build other circuits. One of these basic circuits is the voltage follower. In this cir-

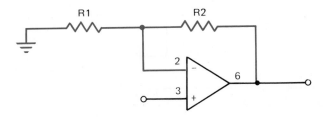

FIGURE 50-5 Noninverted output

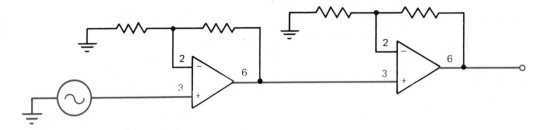

FIGURE 50-6

cuit, the output of the op amp is connected directly back to the inverting input, figure 50-7. Since there is a direct connection between the output of the amplifier and the inverting input, the gain of this circuit is 1. For instance, if a signal voltage of .5 volts is connected to the noninverting input, the output voltage will be .5 volts also. You may wonder why anyone would want an amplifier that does not amplify. Actually, this circuit does amplify something. It amplifies the input impedance by the amount of the open loop gain. If the 741 has an open loop gain of 200,000 and an input impedance of 2 megohms, this circuit would give the amplifier an input impedance of 200K × 2 Meg. or 400,000 megohms. This circuit connection is generally used for impedance matching purposes.

The second basic circuit is the noninverting amplifier, figure 50-8. In this circuit, the output voltage is the same polarity as the input voltage. If the input voltage is a positive-going voltage, the output will be a positive-going voltage at the same time. The amount of gain is set by the ratio of resistors R1 + R2/R1 in the negative feedback loop.

The third basic circuit is the inverting amplifier, figure 50-9. In this circuit the output voltage will be opposite in polarity to the input voltage. If the input signal is a positive-going voltage, the out-

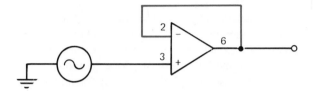

FIGURE 50-7   Voltage follower connection

put voltage will be negative-going at the same instant in time. The gain of the circuit is determined by the ratio of resistors R2 and R1.

CIRCUIT APPLICATIONS

The Level Detector

The operational amplifier is often used as a level detector or comparator. In this type of circuit, the 741 op amp will be used as an inverted amplifier to detect when one voltage becomes greater than another. Refer to the circuit shown in figure 50-10. Notice that this circuit does not use an above-and below-ground power supply. Instead it is connected to a power supply with a single positive and negative output. During normal operation, the non-

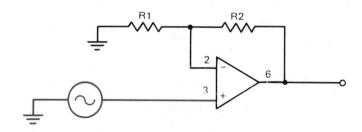

FIGURE 50-8   Noninverting amplifier connection

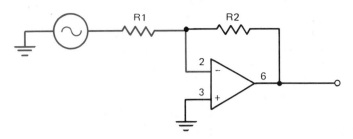

FIGURE 50-9 Inverting amplifier connection

inverting input of the amplifier is connected to a zener diode. This zener diode produces a constant positive voltage at the noninverting input of the amplifier which is used as a reference. As long as the noninverting input is more positive than the inverting input, the output of the amplifier will be high. A light-emitting diode, D1, will be used to detect a change in the polarity of the output. As long as the output of the op amp remains high, the LED will be turned off. When the output of the amplifier is high, the LED has equal voltage applied to both its anode and cathode. Since both the anode and cathode are connected to +12 volts, there is no potential difference and therefore no current flow through the LED.

If the voltage at the inverting input should become more positive than the reference voltage applied to pin #3, the output voltage will go low. The low voltage at the output will be about +2.5 volts. The output voltage of the op amp will not go to 0 or ground in this circuit because the op amp is not connected to a voltage that is below ground. If the output voltage is to be able to go to 0 volts, pin #4 must be connected to a voltage that is below ground. When the output is low there is a potential of about 9.5 volts ($12 - 2.5 = 9.5$) produced across R1 and D1, which causes the LED to turn on and indicate that the state of the op amp's output has changed from high to low.

In this type of circuit, the op amp appears to be a digital device in that the output seems to have only two states, high or low. Actually, the op amp is not a digital device. This circuit only makes it appear digital. Notice there is no negative feedback loop connected between the output and the inverting input. Therefore, the amplifier uses its open loop gain, which is about 200,000 for the 741, to amplify the voltage difference between the inverting input and the noninverting input. If the voltage applied to the inverting input should become one millivolt more positive than the reference voltage applied to the noninverting input, the amplifier will

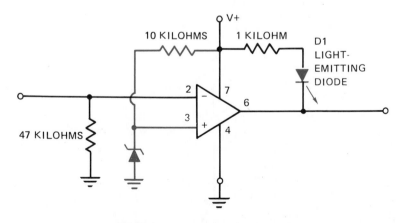

FIGURE 50-10 Inverting level detector

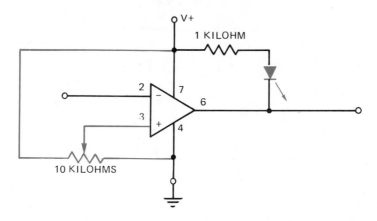

FIGURE 50-11 Adjustable inverting level detector

try to produce an output that is 200 volts more negative than its high-state voltage (.001 × 200,000 = 200). The output voltage of the amplifier cannot be driven 200 volts more negative, of course, because there is only 12 volts applied to the circuit, so the output voltage simply reaches the lowest voltage it can and then goes into saturation. The op amp is not a digital device, but it can be made to act like one.

If the zener diode is replaced with a voltage divider, as shown in figure 50-11, the reference voltage can be set to any value desired. By adjusting the variable resistor shown in figure 50-11, the positive voltage applied to the noninverting input can be set for any voltage value desired. For instance, if the voltage at the noninverting input is set for 3 volts, the output of the op amp will go low when the voltage applied to the inverting input

becomes greater than +3 volts. If the voltage at the noninverting input is set for 8 volts, the output voltage will go low when the voltage applied to the inverting input becomes greater than +8 volts. Notice that this circuit permits the voltage level at which the output of the op amp will change to be adjusted.

In the two circuits just described, the op amp changed from a high level to a low level when activated. There may be occasions, however, when it is desired that the output be changed from a low level to a high level. This can be accomplished by connecting the inverting input to the reference voltage and connecting the noninverting input to the voltage being sensed, figure 50-12. In this circuit, the zener diode is used to provide a positive reference voltage to the inverting input. As long as the voltage at the inverting input remains more

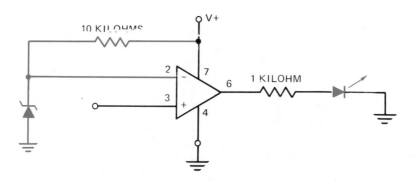

FIGURE 50-12 Noninverting level detector

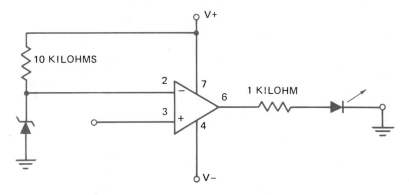

FIGURE 50-13 Below-ground power connection permits the output voltage to become negative.

positive than the voltage at the noninverting input, the output voltage of the op amp will remain low. If the voltage applied to the noninverting input should become more positive than the reference voltage, the output of the op amp will become high.

Depending on the application, this circuit could cause a small problem. As stated previously, since this circuit does not use an above- and below-ground power supply, the low output voltage of the op amp will be about +2.5 volts. This positive output voltage could cause any other devices connected to the op amp's output to be turned on even if it should be turned off. For instance, if the LED shown in figure 50-12 was used, it would glow dimly even when the output is in the low state. This problem can be corrected in a couple of different ways. One way would be to connect the op amp to an above- and below-ground power supply as shown in figure 50-13.

In this circuit, the output voltage of the op amp will be negative or below ground as long as the voltage applied to the inverting input is more positive than the voltage applied to the noninverting input. As long as the output voltage of the op amp is negative with respect to ground, the LED is reverse biased and can not operate. When the voltage applied to the noninverting input becomes more positive than the voltage applied to the inverting input, the output of the op amp will become positive and the LED will turn on.

The second method of correcting the output voltage problem is shown in figure 50-14. In this circuit, the op amp is connected to a power supply that has a single positive and negative output as before. A zener diode, D2, has been connected in series with the output of the op amp and the LED. The voltage value of diode D2 is greater than the output voltage of the op amp in the low state, but

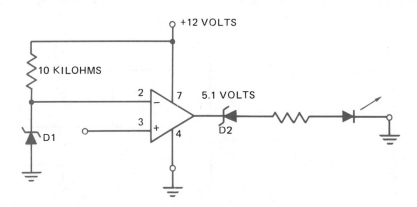

FIGURE 50-14 A zener diode is used to keep the output turned off.

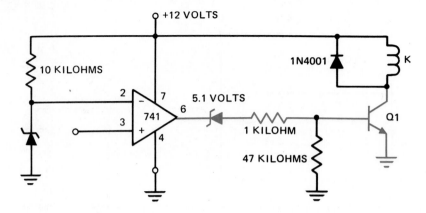

FIGURE 50-15

less than the output voltage of the op amp in its high state. For example, assume the value of the zener diode D2 is 5.1 volts. If the output voltage of the op amp in its low state is 2.5 volts, diode D2 is turned off and will not conduct. If the output voltage becomes +12 volts when the op amp switches to its high state, the zener diode will turn on and conduct current to the LED. Notice that the zener diode D2 keeps the LED turned completely off until the op amp switches to its high state and provides enough voltage to overcome the reverse voltage drop of the zener diode.

In the preceding circuits, an LED was used to indicate the output state of the amplifier. Keep in mind that the LED is used only as a detector, and the output of the op amp could be used to control almost anything. For example, the output of the op amp can be connected to the base of a transistor as shown in figure 50-15. The transistor can then control the coil of a relay, which could be used to control almost anything.

The Oscillator

An operational amplifier can be used as an oscillator. The circuit shown in figure 50-16 is a very simple circuit which will produce a square wave output. This circuit is rather impractical, however. This circuit would depend on a slight imbalance in the op amp or random circuit noise to start the oscillator. A slight voltage difference of a few millivolts between the two inputs is all that is needed to cause the output of the amplifier to go high or low. For example, if the inverting input becomes slightly more positive than the noninverting input, the output will go low or negative. When the output becomes negative, capacitor Ct begins to charge through resistor Rt to the negative value of the out-

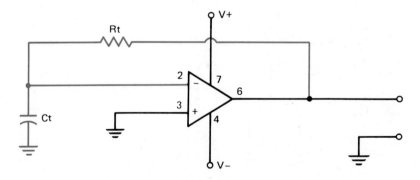

FIGURE 50-16 Simple square wave oscillator

put voltage. As soon as the voltage applied to the inverting input becomes slightly more negative than the voltage applied to the noninverting input, the output will change to a high or positive value of voltage. When the output becomes positive, capacitor Ct begins charging through resistor Rt toward the positive output voltage. This circuit will work quite well if the op amp has no imbalance, and if the op amp is shielded from all electrical noise. In practical application, however, there is generally enough imbalance in the amplifier or enough electrical noise to send the op amp into saturation, which stops the operation of the circuit.

The Hysteresis Loop

The problem with this circuit is that a millivolt difference between the two inputs is enough to drive the amplifier's output from one state to the other. This problem can be corrected by the addition of a hysteresis loop connected to the noninverting input as shown in figure 50-17. Resistors R1 and R2 form a voltage divider for the noninverting input. These resistors are generally of equal value. To understand the circuit operation, assume that the inverting input is slightly more positive than the noninverting input. This causes the output voltage to go negative. Also assume that the output voltage is now negative 12 volts as compared to ground. If resistors R1 and R2 are of equal value, the noninverting input is driven to −6 volts by the voltage divider. Capacitor Ct begins to charge through resistor Rt to the value of the output voltage. When capacitor Ct has been charged to a value slightly

more negative than the −6 volts applied to the noninverting input, the op amp's output goes high or to +12 volts above ground. When the output of the op amp changes from −12 volts to +12 volts, the voltage applied to the noninverting input changes from −6 volts to +6 volts. Capacitor Ct now begins to charge through resistor Rt to the positive voltage of the output. When the voltage applied to the inverting input becomes more positive than the voltage applied to the noninverting input, the output changes to a low value or −12 volts. The voltage applied to the noninverting input is driven from +6 volts to −6 volts, and capacitor Ct again begins to charge toward the negative output voltage of the op amp. Notice that the addition of the hysteresis loop has greatly changed the operation of the circuit. The voltage differential between the two inputs is now volts instead of millivolts. The output frequency of the oscillator is determined by the values of Ct and Rt. The period of one cycle can be computed by using the formula ($T = 2RC$).

The Pulse Generator

The operational amplifier can also be used as a pulse generator. The difference between an oscillator and a pulse generator is the period of time the output remains on as compared to the period of time it remains low or off. An oscillator is generally considered to produce a waveform that has positive and negative pulses of equal voltage and time, figure 50-18. Notice that the positive value of voltage is the same as the negative value. Also notice that both the positive and negative cycles

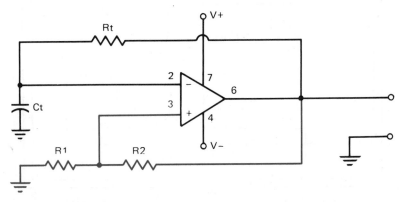

FIGURE 50-17 Square wave oscillator using a hysteresis loop

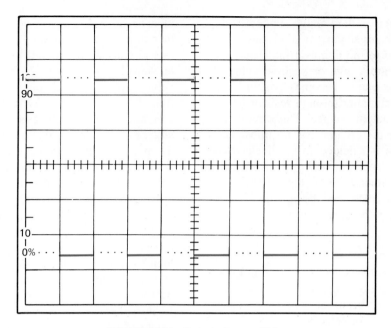

FIGURE 50-18 Output of an oscillator

remain turned on the same amount of time. This waveform is consistent with that which one would expect to see if an oscilloscope is connected to the output of a square wave oscillator.

If the oscilloscope is connected to a pulse generator, however, a waveform similar to the one shown in figure 50-19 would be seen. Notice that the positive value of voltage is the same as the neg-

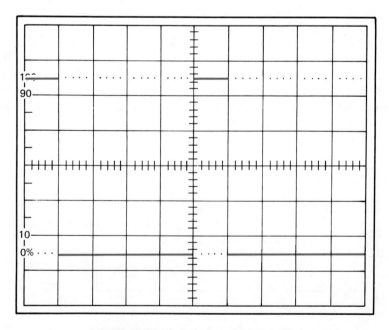

FIGURE 50-19 Output of a pulse generator

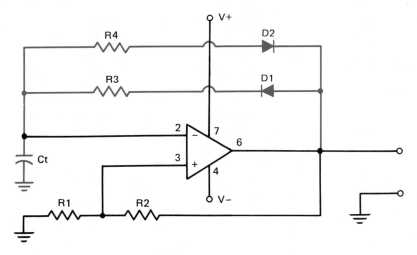

FIGURE 50-20 Pulse generator circuit

ative value just as it was in figure 50-18. However, the positive pulse is of a much shorter duration than the negative pulse. The device producing this waveform is generally considered to be a pulse generator rather than an oscillator.

The 741 operational amplifier can easily be changed from a square wave oscillator to a pulse generator. Refer to the circuit shown in figure 26-20. This is the same basic circuit as the square wave oscillator with the addition of resistor R3 and R4, and diodes D1 and D2. This circuit permits capacitor Ct to charge at a different rate when the output is high or positive than it does when the output is low or negative. For instance, assume that the voltage of the op amp's output is low or −12 volts. If the output voltage is negative, diode D1 is reverse biased and no current can flow through resistor R3. Therefore, capacitor Ct must charge through resistor R4 and diode D2, which is forward biased. When the voltage applied to the inverting input becomes more negative than the voltage applied to the non-inverting input, the output voltage of the op amp becomes +12 volts. When the output voltage becomes +12 volts, diode D2 is reverse biased and

FIGURE 50-21 741 operational amplifier in an eight-pin in-line case

diode D1 is forward biased. Capacitor Ct, therefore, begins charging toward the +12 volts through resistor R3 and diode D1. Notice that the amount of time the output of the op amp remains low is determined by the value of Ct and R4, and the amount of time the output remains high is determined by the value of Ct and R3. The ratio of time the output voltage is high compared to the amount of time it is low can be determined by the ratio of resistor R3 to resistor R4. A 741 operational amplifier is shown in figure 50-21.

REVIEW QUESTIONS

1. When the voltage connected to the inverting input is more positive than the voltage connected to the noninverting input, will the output be positive or negative?

2. What is the input impedance of a 741 operational amplifier?

3. What is the average open loop gain of the 741 operational amplifier?

4. What is the average output impedance of the 741?

5. List the three common connections for operational amplifiers.

6. When the operational amplifier is connected as a voltage follower, it has a gain of one. If the input voltage does not get amplified, what does?

7. Name two effects of negative feedback.

8. Refer to figure 50-8. If resistor R1 is 200 ohms, the resistor R2 is 10K ohms, what is the gain of the amplifier?

9. Refer to figure 50-9. If resistor R1 is 470 ohms and resistor R2 is 47K ohms, what is the gain of the amplifier?

10. What is the purpose of the hysteresis loop when the op amp is used as an oscillator?

SECTION 9

Solid-State Controls

UNIT 51

Programmable Controllers

Programmable controllers were first used by the automobile industry in the late 1960s. Each time a change in design was made, it was necessary to change the control systems operating the machinery. This consisted of physically rewiring the control system to make it perform the new operation. This, of course, was extremely time consuming and costly. What the industry needed was some type of control system that could be changed without the extensive rewiring required to change relay control systems.

One of the first questions that is generally asked is, "Is a programmable controller a computer?" The answer to that question is "yes." The programmable controller, PC, is a very special computer designed to perform a special function.

PC AND COMMON COMPUTER DIFFERENCES

Some differences between a PC and a home and business computer are:

1. The PC is designed to be operated in an industrial environment. Any computer used in industry must be able to operate in extremes of temperature; ignore voltage spikes and drops on the incoming power line; survive in an atmosphere that often contains corrosive vapors, oil, and dirt; and withstand shock and vibration.
2. Most programmable controllers are designed to

be programmed with relay schematic and ladder diagrams instead of the common computer languages such as BASIC or Fortran. An electrician who is familiar with relay-logic diagrams can generally be trained to program a PC in a few hours. It generally requires several months to train someone to program a standard computer.

BASIC COMPONENTS

Programmable controllers can be divided into four basic parts:

1. The power supply,
2. The central processing unit,
3. The program loader or terminal,
4. The I/O (pronounced eye-oh) track.

THE POWER SUPPLY

The power supply is used to lower the incoming AC voltage to the desired level, rectify it to DC, and then filter and regulate it. The internal logic circuits of programmable controllers operate on 5 to 15 volts DC, depending on the type of controller. This voltage must be free of voltage spikes and other electrical noise. It must also be regulated to within 5% of the required voltage value. Some manufacturers of PCs use a separate power supply,

FIGURE 51-1 Internal components of programmable controller (Courtesy of Struthers-Dunn Inc.)

and others build the power supply into the central processor.

THE CPU

The central processor unit, CPU, is the brains of the programmable controller. It contains the microprocessor chip and related integrated circuits to perform all the logic functions. The microprocessor chip used in most PCs is a common computer chip used in many home and business machines, figure 51-1.

The central processor unit generally has a key switch located on the front panel. This switch must be turned on before the CPU can be programmed. This is done to prevent the circuit from being changed accidently. Plug connections mounted on the central processor are used to provide connection for the programming terminal and the I/O tracks, figure 51-2. Most CPUs are designed so that once the program has been tested, it can be stored on tape or disc. In this way if a central processor

FIGURE 51-2 Central processor unit (Courtesy of Struthers-Dunn Inc.)

unit should fail and have to be replaced, the new unit can be reprogrammed from the tape or disc. This eliminates the time-consuming process of having to reprogram by hand.

THE PROGRAMMING TERMINAL

The programming terminal or loading terminal is used to program the CPU. Most terminals are one of two types. One type is a small handheld device that uses a liquid crystal display to show the program. This terminal, however, will display only one line of the program at a time.

The other type of terminal uses a cathode ray tube, CRT, to show the program. This terminal generally looks similar to a portable television set with a keyboard attached, figure 51-3. They will generally display from four to six lines of the program at a time, depending on the manufacturer.

The terminal is not only used to program the controller, but it is also used to troubleshoot the circuit. When the terminal is connected to the CPU, the circuit can be examined while it is in operation. Figure 51-4 illustrates a typical circuit that could be seen on the display. Notice that this schematic diagram is a little different from the typical ladder diagram. All line components are shown as normally open or normally closed contacts. There are no NEMA symbols of push buttons, float switches, limit switches, etc. The programmable controller recognizes only open or closed contacts. It does

FIGURE 51-3 Programming terminal (Courtesy of General Electric Co.)

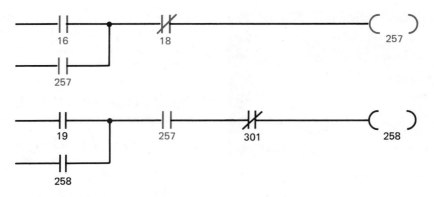

FIGURE 51-4 Analyzing circuit operation within the terminal

not know if a contact is controlled by a push button, a limit switch, or a float switch. Each contact, however, does have a number. The number is used to distinguish one contact from another. The coil symbols look like a set of parentheses instead of a circle, as shown on most ladder diagrams. Each line ends with a coil, and each coil has a number. When a contact symbol has the same number as a coil, it means the contact is controlled by that coil. Figure 51-4 shows a coil numbered 257, and two contacts numbered 257. When relay coil 257 is energized, the controller interprets both of these contacts to be closed.

Notice that both 257 contacts, contacts 16 and 18, and coil 257 are shown with dark heavy lines. When a contact has a complete circuit through it, or a coil is energized, the terminal will illuminate that contact or coil. Contact 16 is illuminated, which means that it is closed and provides a current path. Contact 18 is closed and provides a current path to coil 257. Since coil 257 is energized, both 257 contacts are closed and provide a current path.

Contacts 19, 258, and 301 are not shown illuminated. This means that contacts 19 and 258 are de-energized and still open. Contact 301, however, has been energized. This contact is shown as normally closed. Since it is not illuminated, it is open and no current path exists through it. Notice that when a contact is illuminated, it does not mean the contact has energized or changed position. It means there is a complete path for current flow.

When the terminal is used to load a program into the central processor unit, contact and coil

symbols on the keyboard are used. These symbol keys are used to load a ladder diagram similar to figure 51-4 into the CPU. Programming will be discussed in unit 52.

THE I/O TRACK

The I/O track is used to connect the central processor unit to the outside world. It contains input modules that carry information to the CPU, and output modules that carry information from the CPU. An I/O track with input and output modules is shown in figure 51-5. Most modules contain more than one input or output. Any number from two to eight is common depending on the manufacturer. The modules shown in figure 51-5 can each handle four connections. This means that each input module

FIGURE 51-5 I/O track with input and output modules (Courtesy of Struthers-Dunn Inc.)

can handle four different inputs from pilot devices such as pushbuttons, float switches, or limit switches. Each output module can control four external devices such as pilot lights, solenoids, or motor starter coils. The operating voltage of modules can be AC or DC and are generally either 120 volts or 24 volts. The I/O track in figure 51-5 can handle eight modules. Since each module can handle four devices, this I/O track can handle 32 inputs or outputs.

I/O CAPACITY

One factor that determines the size and cost of a programmable controller is its I/O capacity. Many small units are designed to handle only 32 inputs or outputs. Large units can handle several hundred. The controller shown in figure 51-2 is designed to handle eight I/O tracks. Since each I/O track has 32 inputs or outputs, the controller has an I/O capacity of 256.

INPUT MODULES

The central processor unit of a programmable controller is extremely sensitive to voltage spikes and electrical noise. For this reason the input I/O uses optoisolation to electrically separate the incoming signal from the CPU.

Figure 51-6 shows a typical circuit used for the input. The bridge rectifier changes the AC volt-age into DC. A resistor is used to limit current to the light emitting diode (LED). When the LED turns on, the light is detected by the phototransistor, which signals the CPU that there is a voltage present at the input terminal.

When the module has more than one input, one side of the bridge rectifiers is connected together to form a common terminal. The other side of the rectifiers is labelled 1, 2, 3, and 4. Figure 51-7 shows four bridge rectifiers connected together to form a common terminal.

Figure 51-8 shows a limit switch connected to the input. Notice that the limit switch completes a circuit from the AC line to the bridge rectifier. When the limit switch closes, 120 VAC is applied to the rectifier and the LED turns on.

THE OUTPUT MODULE

The output module is used to provide connection from the central processor unit to the load. The output is an optoisolated solid-state relay. The current rating can range from .5 to 3 amps, depending on the manufacturer. Voltage ratings are generally 24 or 120 volts and can be AC or DC.

If the output is designed to control a DC voltage, a power transistor is used to control the load, figure 51-9. The transistor is a phototransistor that is operated by a light-emitting diode. The LED is operated by the CPU.

If the output is designed to control an AC load,

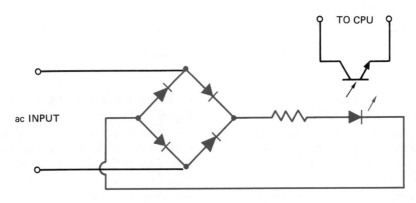

FIGURE 51-6   Input circuit

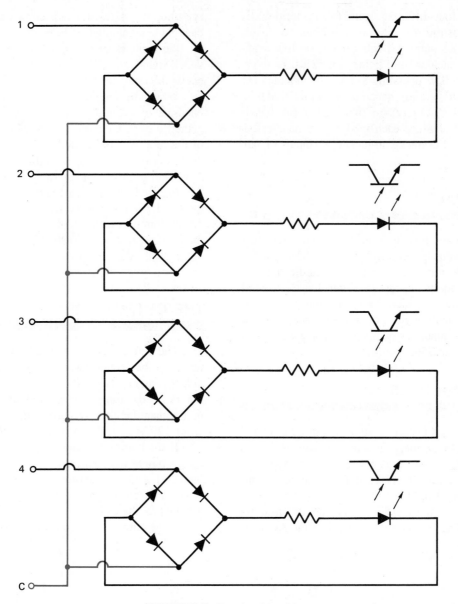

FIGURE 51-7 Four input modules

a triac is used as the control device instead of a power transistor, figure 51-10. A photodetector connected to the gate of the triac is used to control the output. When the LED is turned on by the CPU, the photodetector permits current to flow through the gate of the triac and turn it on.

If more than one output is contained in a module, one side of the control devices are generally connected together to form a common terminal. Figure 51-11 shows an output module that contains four outputs. Notice that one side of each triac has been connected to form a common terminal. The other side of the triacs are labelled 1, 2, 3, and 4. If power transistors are used as the control devices, the emitters or the collectors can be connected to form a common terminal.

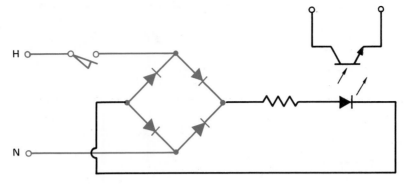

FIGURE 51-8 Limit switch completes circuit to rectifier.

Figure 51-12 shows a solenoid coil connected to an AC output module. Notice that the triac is used as a switch to complete a circuit so that current can flow through the coil. The output module does not provide power to operate the load. The power must be provided by an external power source. The amount of current an output can control is limited. Small current loads such as solenoid coils and pilot lights can be controlled directly by the I/O output, but large current loads, such as motors, cannot. When a large amount of current must be controlled, the output is used to operate the coil of a motor starter or contactor. The motor starter or contactor can be used to control almost anything.

INTERNAL RELAYS

The actual logic of the control circuit is performed by internal relays. An internal relay is an imaginary device that exists only in the logic of the computer. They can have any number of contacts

from one to several hundred, and the contacts can be normally open or normally closed. Internal relays can be programmed into the computer by assigning a coil some number greater than the I/O capacity. For example, assume the programmable controller has an I/O capacity of 256. If a coil is programmed into the computer and assigned a number greater than 256, 257 for instance, it is an internal relay. Any number of contacts can be controlled by relay 257 by simply inserting a contact symbol in the program and numbering it 257. If a coil is numbered 256 or less, it can turn on an output when energized.

Inputs are programmed in a similar manner. If a contact is inserted in the program and assigned a number less than 256, the contact will be changed when a voltage is sensed at that input point. For example, assume a normally open contact has been programmed in the circuit and assigned number 22. When voltage is applied to input number 22, the contact will close. Since 22 is used as an input in this circuit, care must be taken not to assign number 22 to a coil. Terminal number 22 cannot be

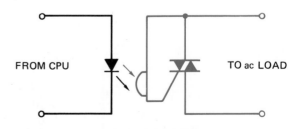

FIGURE 51-9 Output used to control DC

FIGURE 51-10 Output module used to control an AC module

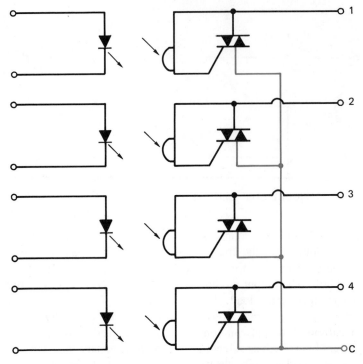

FIGURE 51-11 Four AC outputs in one module

COUNTERS AND TIMERS

The internal relays of a programmable controller can be used as counters and timers. When timers are used, most of them are programmed in .1 second intervals. For example, assume a timer is to be used to provide a delay of 10 seconds. When the delay time is assigned to the timer, the number 00100 would be used. This means the timer has been set for 100 tenths of a second, which is 10 seconds.

OFF-DELAY CIRCUIT

The internal timers of a programmable controller function as on-delay relays. A simple circuit can be used, however, to change the sense of the on-delay timer to make it perform as an off-delay timer. Figure 51-13 is this type of circuit. The de-

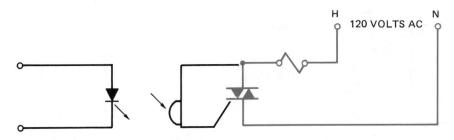

FIGURE 51-12 An output controls a solenoid.

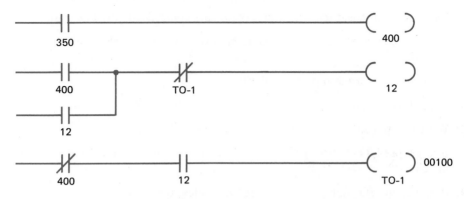

FIGURE 51-13 Off-delay circuit

sired operation of the circuit is as follows: When contact 350 closes, relay coil 12 energizes immediately and turns on a solenoid valve. When contact 350 opens, coil 12 will remain energized for 10 seconds before it de-energizes and turns off the solenoid.

This logic is accomplished as follows:

1. When contact 350 closes, internal relay 400 energizes.
2. When coil 400 energizes, normally open contact 400 closes and completes a circuit to coil 12, and the normally closed 400 contact connected in series with timer TO-1 opens.
3. When relay coil 12 energizes, both normally open 12 contacts close, and the I/O output at terminal 12 connects the solenoid coil to the power line.
4. When contact 350 opens, internal relay 400 de-energizes.
5. This causes both 400 contacts to change back to their original position.
6. When the normally open 400 contact returns to its open state, a continued current path to coil 12 is maintained by the now closed 12 contact connected in parallel with it.
7. When the normally closed 400 contact returns

to its closed position, a circuit is completed through the now closed 12 contact to TO-1 coil.
8. When TO-1 coil is energized, a 10-second timer is started. At the end of this time period, contact TO-1 opens and de-energizes coil 12.
9. When coil 12 de-energizes, both 12 contacts return to their open position and the output I/O turns the solenoid off.
10. TO-1 timer de-energizes when 12 contact opens and the circuit is back in its original start condition.

The number of internal relays and timers contained in a programmable controller is determined by the memory capacity of the computer. As a general rule, PCs that have a large I/O capacity will have a large memory, and machines with less I/O capacity have less memory.

The use of programmable controllers has steadily increased since their invention in the late 1960s. A PC can replace hundreds of relays and occupy only a fraction of the space. The circuit logic can be changed easily and quickly, without the need of extensive hand rewiring. They have no moving parts or contacts to wear out, and their down time is less than an equivalent relay circuit.

REVIEW QUESTIONS

1. What industry first started using programmable controllers?
2. Name two differences between PCs and home or business computers.

3. Name the four basic sections a programmable controller is divided into.

4. In what section of the programmable controller is the actual circuit logic performed?

5. What device is used to program the PC?

6. What device separates the programmable controller from the outside circuits?

7. What two functions are performed by an input I/O?

8. If an output I/O controls DC voltage, what electronic device is used to control the circuit?

9. If an output I/O controls AC voltage, what electronic device is used to control the circuit?

10. What is an internal relay?

UNIT 52
Programming a PC

In this unit a relay schematic will be converted into a diagram used to program a programmable controller. The process to be controlled is shown in figure 52-1. Although this is not an air-conditioning circuit, it was chosen because it is simple and uses different functions of the programmable controller. In this circuit, a tank is used to mix two liquids. The operation of the control circuit is as follows:

1. When a start button is pressed, solenoids A and B energize. This permits the two liquids to begin filling the tank.

2. When the tank is filled, the float switch trips. This de-energizes solenoids A and B and starts the motor used to mix the liquids together.

3. The motor is permitted to run for one minute. At the end of this time, the motor turns off and solenoid C energizes to drain the tank.

4. When the tank is empty, the float switch de-energizes solenoid C.

5. A stop button can be used to stop the process at any point.

6. If the motor should become overloaded, the action of the entire circuit will stop.

7. Once the circuit has been energized, it will continue to operate until it is manually stopped.

CIRCUIT OPERATION

A relay schematic that will perform the logic of this circuit is shown in figure 52-2. The logic of this circuit is as follows:

1. When the start button is pushed, CR relay coil is energized. This causes all CR contacts to close. CR-1 contact is a holding contact used to maintain the circuit to CR coil when the start button is released.

2. When CR-2 contact closed, a circuit is completed to solenoid coils A and B. This permits the two liquids to be mixed together to begin filling the tank.

3. As the tank fills, the float rises until the float

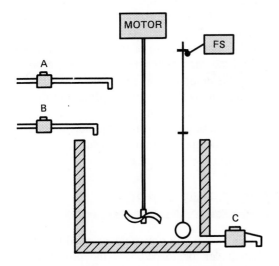

FIGURE 52-1 Tank used to mix two liquids

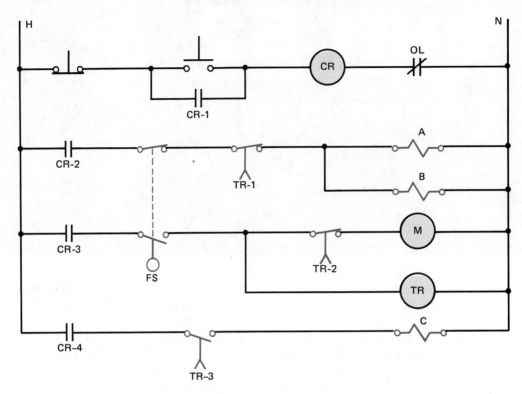

FIGURE 52-2 Relay schematic

switch is tripped. This causes the normally closed float switch contact to open and the normally open contact to close.

4. When the normally closed float switch opens, solenoid coils A and B de-energize and stop the flow of the two liquids into the tank.

5. When the normally open contact closes, a circuit is completed to the coil of a motor starter and the coil of an on delay timer. The motor is used to mix the two liquids together.

6. At the end of a 1-minute time period, all of the TR contacts change position. The normally closed TR-2 contact connected in series with the motor starter coil opens and stops the operation of the motor. The normally open TR-3 contact closes and energizes solenoid coil C, which permits liquid to begin draining from the tank. The normally closed TR-1 contact is used to ensure valves A and B cannot be re-energized until solenoid C de-energizes.

7. As liquid drains from the tank, the float drops. When the float drops far enough, the float switch

trips and its contacts return to their normal position. When the normally open float switch contact reopens and de-energizes TR coil, all TR contacts return to their normal positions.

8. When the normally open TR-3 contact reopens, solenoid C de-energizes and closes the drain valve. TR-2 contact recloses, but the motor cannot restart because of the normally open float switch contact. When TR-1 contact recloses, a circuit is completed to solenoids A and B. This permits the tank to begin refilling and the process starts over again.

9. If the stop button or overload contact opens, CR coil de-energizes and all CR contacts open. This de-energizes the entire circuit.

DEVELOPING A PROGRAM

This circuit will now be developed into a program that can be loaded into the programmable controller. It will be assumed that the controller to be used has an I/O capacity of 32. It is also assumed that I/O terminals 1 through 16 are used as inputs, and terminals 17 through 32 are used as outputs.

The first line of the relay schematic in figure 52-2 shows a normally closed stop button connected in series with a normally open start button. A normally open CR contact is connected in parallel with the start button, and CR coil and the normally closed overload contact are connected in series with the start button.

Figure 52-3 shows lines 1 and 2 of the program that will be used to program the PC. Contact #1 is controlled by the normally closed stop push button connected to terminal #1 of the I/O track. Notice that contact #1 is programmed normally open instead of normally closed. Since the stop button is a normally closed push button, power will be applied to terminal #1 of the I/O track during the normal operation of the circuit. When power is applied to an input terminal of the I/O track, the central processor unit interprets this as an instruction to change the position of the contact that corresponds to that terminal number. Therefore, the CPU will interpret contact #1 to be closed during the normal operation of the circuit.

Contact #2 is connected to the normally open start button. Contact #3 is connected to a normally closed overload contact. Notice that this contact is programmed normally open for the same reason that contact #1 is programmed normally open. Notice also that the position of the overload contact has been changed from the right side of the coil to the left side. This is done because a coil ends the line of a program. All contacts used to control a coil must be placed ahead of it. The coil is assigned #33. Since this programmable controller has an I/O capacity of 32, coil 33 is an internal relay.

The normally open 33 contact is connected in parallel with contact #2. Although this contact is the only component on a line, it counts as one full line of the program.

Figure 52-4 shows the addition of two more lines of the program. Contact #4 is connected to the float switch. The float switch controls the operation of internal relay 34. Internal relay 35 is controlled by contacts 33, 34, and TO-1.

Figure 52-5 shows the addition of two more lines of the program. When internal relay 35 is energized, outputs 17 and 18 turn on. Solenoid coil A is connected to terminal 17 and the I/O track and solenoid coil B is connected to terminal 18.

Figure 52-6 shows the complete program. Coil 19 controls output 19, which is connected to the coil of a motor starter. Coil TO-1 is an internal timer that has been programmed for 600 tenths of a second, which is 60 seconds. Coil 20 controls output 20, which is connected to solenoid coil C.

OPERATION OF THE PC CIRCUIT

The operation of this circuit is as follows:

1. Since the stop button and overload contact are normally closed, inputs #1 and #3 have continuous voltage applied to them. The central processor unit, therefore, interprets contacts #1 and #3 to be closed.
2. When the start button is pushed, contact #2 closes. This completes a circuit to coil 33.
3. When coil 33 energizes, all 33 contacts close. The start button can now be released and the circuit will be maintained by the 33 contact connected in parallel with contact #2. Internal relay coil 35 energizes when a 33 contact closes.
4. When contacts 35 close, coils 17 and 18 energize. This permits solenoids A and B to energize and begin filling the tank with liquid.

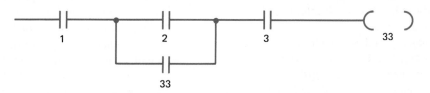

FIGURE 52-3 Lines 1 and 2 of the program

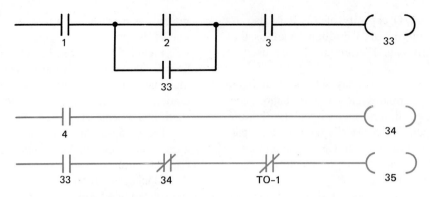

FIGURE 52-4 Two more lines are added to the program.

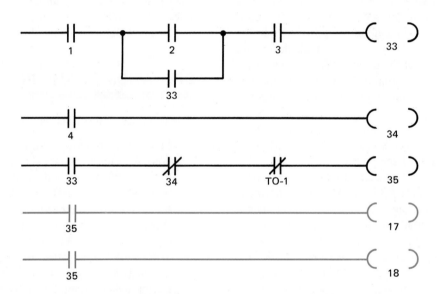

FIGURE 52-5 Solenoids A and B are connected to outputs 17 and 18.

5. When the float rises high enough, the float switch activates and supplies power to terminal #4 of the I/O track. The CPU interprets contact #4 to be closed and internal relay coil 34 energizes.

6. The normally closed 34 contact opens and de-energizes coil 35. This causes both of the 35 contacts to open and de-energize coils 17 and 18. When output terminal 17 and 18 turn off, solenoids A and B de-energize.

When the normally open 34 contact closes, coil 19 energizes, which connects power to the motor starter coil, and TO-1 timer coil energizes and starts the 60-second timer.

7. At the end of the 60-second time period, all TO-1 contacts change position. One of the normally closed contacts opens and prevents coil 35 from energizing until the timer has reset. The other normally closed contact opens and de-energizes output 19 which de-energizes

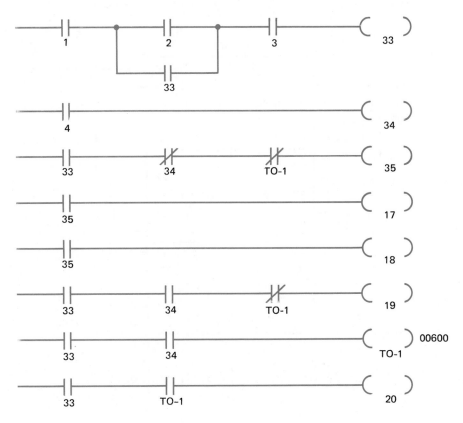

FIGURE 52-6 Complete program

the motor starter coil to stop the mixing motor. The normally open TO-1 contact closes and energizes output 20. This output energizes solenoid C, which begins draining the tank. As liquid drains from the tank, the float drops. When the float drops far enough, the float switch opens and breaks the circuit to input terminal #4. This causes contact #4 to open and de-energize coil 34.

8. When coil 34 de-energizes, all 34 contacts return to their normal position. This causes TO-1 coil to de-energize and all TO-1 contacts return to their normal position.

9. This de-energizes coil 20, which turns off solenoid C, and also completes the circuit to coil 35. When coil 35 energizes, outputs 17 and 18 turn on and energize solenoids A and B to start the process over again.

10. If the stop pushbutton should be pushed or if the overload contact should open, coil 33 will de-energize and stop the operation of the circuit.

Notice that this circuit operates with the same logic as the relay circuit. The circuit in figure 52-6 is now ready to be programmed into the controller. Pushbuttons located on the programming terminal, which represent open and closed contacts, coils, and timers are used to insert the program into the central processor unit.

Figure 52-7 shows the connection of external components to the I/O track. The pushbuttons, overload contact, and float switch are connected to the input terminals. The solenoid coils and motor starter coil are connected to the output terminals.

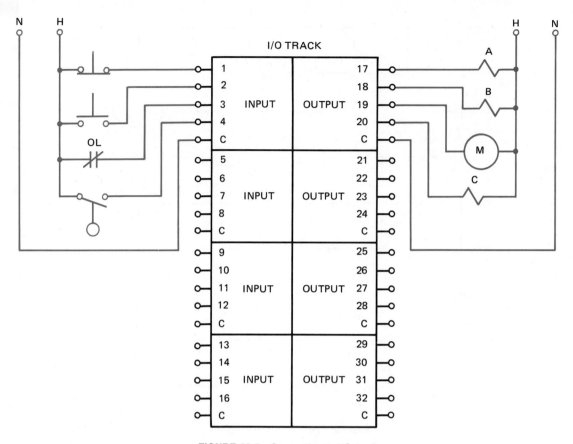

FIGURE 52-7 Connection to I/O track

REVIEW QUESTIONS

1. Why are NEMA symbols such as push buttons, limit switches, and float switches not used in a programmable-controller schematic?

2. Explain how to program an internal relay into the controller.

3. Why are contacts used to represent stop buttons and overload contacts programmed normally open?

4. Why is the output I/O used to energize a motor starter instead of energizing the motor directly?

5. A timer is to be programmed for a delay of 3 minutes. What number is used to set this timer?

UNIT 53

Analog Sensing for Programmable Controllers

Many of the programmable controllers found in industry are designed to accept analog as well as digital inputs. Analog means continuously varying. These inputs are designed to sense voltage, current, speed, pressure, proximity, temperature, etc. When an analog input is used, such as a thermocouple for measuring temperature, a special module that mounts on the I/O track is used. These types of sensors are often used with set point detectors which can be used to trigger alarms and turn on or off certain processes. For example, the voltage produced by a thermocouple will increase with a change of temperature. Assume it is desired to sound an alarm if the temperature of an object reaches a certain level. The detector is preset with a particular voltage. As the temperature of the thermocouple increases, its output voltage increases also. When the voltage of the thermocouple becomes greater than the preset voltage, an alarm sounds.

INSTALLATION

Most analog sensors can produce only very weak signals. Zero to 10 volts or 4 to 20 milliamps is common. In an industrial environment where intense magnetic fields and large voltage spikes abound, it is easy to lose the input signal amid the electrical noise. For this reason, special precautions should be taken when installing the signal wiring between the sensor and input module. These precautions are particularly important when using analog inputs, but they should be followed when using digital inputs also.

KEEP WIRE RUNS SHORT

Try to keep wire runs as short as possible. The longer a wire run is, the more surface area of wire there is to pick up stray electrical noise.

PLAN THE ROUTE OF THE SIGNAL CABLE

Before starting, plan how the signal cable should be installed. Never run signal wire in the same conduit with power wiring. Try to run signal wiring as far away from power wiring as possible. When it is necessary to cross power wiring, install the signal cable so that is crosses at a right angle as shown in figure 53-1.

USE SHIELDED CABLE

Shielded cable is generally used for the installation of signal wiring. One of the most common types, figure 53-2, uses twisted wires with a mylar foil shield. The ground wire must be grounded if the shielding is to operate properly. This type of

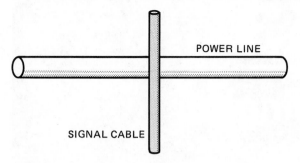

FIGURE 53-1 Signal cable crosses power line at right angle.

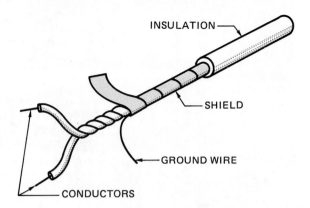

FIGURE 53-2 Shielded cable

shielded cable can provide a noise reduction ratio of about 30,000:1.

Another type of signal cable uses a twisted pair of signal wires surrounded by a braided shield. This type of cable provides a noise reduction of about 300:1.

Common co-axial cable should be avoided. This cable consists of a single conductor surrounded by a braided shield. This type of cable offers very poor noise reduction.

GROUNDING

Ground is generally thought of as being electrically neutral or zero at all points. This may not be the case in practical application, however. It is not uncommon to find different pieces of equipment that have ground levels that are several volts apart, figure 53-3.

One method that is sometimes used to overcome this problem is to use large cable to tie the two pieces of equipment together. This forces them to exist at the same potential. This method is some-

times referred to as the brute force method.

Where the brute force method is not practical, the shield of the signal cable is grounded at only one end. The preferred method is generally to ground the shield at the sensor.

THE DIFFERENTIAL AMPLIFIER

An electronic device that is often used to help overcome the problem of induced noise is the differential amplifier shown in figure 53-4. This device detects the voltage difference between the pair of signal wires and amplifies this difference. Since the induced noise level should be the same in both conductors, the amplifier will ignore the noise. For example, assume an analog sensor is producing a 50 millivolt signal. This signal is applied to the input module, but induced noise is at a level of 5 volts. In this case the noise level is 100 times greater than the signal level. The induced noise level,

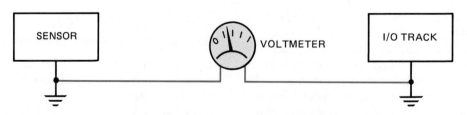

FIGURE 53-3 All grounds are not equal.

SECTION 9 SOLID-STATE CONTROLS

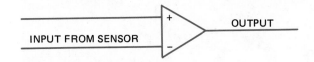

INPUT FROM SENSOR · OUTPUT

FIGURE 53-4 Differential amplifier detects difference of signal level.

however, is the same for both of the input conductors. The differential amplifier, therefore, ignores the 5-volt noise and amplifies only the voltage difference which is the 50 millivolts.

REVIEW QUESTIONS

1. Explain the difference between digital inputs and analog inputs.

2. Why should signal-wire runs be kept as short as possible?

3. When signal wiring must cross power wiring, how should the crossing be done?

4. Why is shielded wire used for signal runs?

5. What is the brute force method of grounding?

6. Explain the operation of the differential amplifier.

GLOSSARY

AC (alternating current) Current that reverses its direction of flow periodically. Reversals generally occur at regular intervals.

ACROSS-THE-LINE A method of motor starting that connects the motor directly to the supply line on starting or running. (Also known as Full Voltage Starting.)

ALTERNATOR A machine used to generate alternating current by rotating conductors through a magnetic field.

AMBIENT TEMPERATURE The temperature in the surrounding area of a device.

AMPACITY The maximum current rating of a wire or device.

AMPLIFIER A device used to increase a signal.

AMPLITUDE The highest value reached by a signal, voltage, or current.

ANALOG VOLTMETER A voltmeter that uses a meter movement to indicate the voltage value.

ANODE The positive terminal of an electrical device.

ANTI-SHORT-CYCLING A control that prevents the compressor from being restarted within a certain time after it has stopped.

APPLIED VOLTAGE The amount of voltage connected to a circuit or device.

ASA American Standards Association.

ATOM The smallest part of an element that contains all the properties of that element.

ATTENUATOR A device that decreases the amount of signal voltage or current.

AUTOMATIC Self-acting, operation by its own mechanical or electrical mechanism.

AUXILIARY CONTACTS Small contacts located on relays and motor starters used for the purpose of operating other control components.

BASE The semi-conductor region between the collector and emitter of a transistor. The base controls the current flow through the collector-emitter circuit.

BIAS A DC voltage applied to the base of a transistor to preset its operating point.

BIMETAL STRIP A strip made by bonding two unlike metals together that, when heated, expand at different temperatures. This causes a bending or warping action.

BRANCH CIRCUIT That portion of a wiring system that extends beyond the circuit protective device such as a fuse or circuit breaker.

BREAKDOWN TORQUE The maximum amount of torque that can be developed by a motor at rated voltage and frequency before an abrupt change in speed occurs.

BRIDGE CIRCUIT A circuit that consists of four sections connected in series to form a closed loop.

BRIDGE RECTIFIER A device constructed with four diodes that converts both positive and negative cycles of AC voltage into DC voltage. The bridge rectifier is one type of full-wave rectifier.

BUSWAY An enclosed system used for power transmission that is voltage and current rated.

CAPACITANCE The electrical size of a capacitor.

CAPACITOR A device made with two conductive plates separated by an insulator or dielectric.

CAPACITOR-START MOTOR A single-phase induction motor that uses a capacitor connected in series with the start winding to increase starting torque.

CATHODE The negative terminal of an electrical device.

CENTER-TAPPED TRANSFORMER A

transformer that has a wire connected to the electrical midpoint of its winding. Generally, the secondary winding is tapped.

CHOKE An inductor designed to present an impedance to AC current, or to be used as the current filter of a DC power supply.

CIRCUIT BREAKER A device designed to open under an abnormal amount of current flow. The device is not damaged and may be used repeatedly. They are rated by voltage, current, and horsepower.

CLOCK TIMER A time-delay device that uses an electric clock to measure the delay time.

COLLAPSE (of a magnetic field) Occurs when a magnetic field suddenly changes from its maximum value to a zero value.

COLLECTOR A semi-conductor region of a transistor that must be connected to the same polarity as the base.

COMPARATOR A device or circuit that compares two like quantities, such as voltage levels.

COMPRESSOR The component of an air-conditioning or refrigeration system that maintains the difference in pressure between the high and low sides.

CONDENSING UNIT The component of an air-conditioning or refrigeration system in which heat is removed from the refrigerant and dissipated to the surrounding air or liquid.

CONDUCTION LEVEL The point at which an amount of voltage or current will cause a device to conduct.

CONDUCTOR A device or material that permits current to flow through it easily.

CONTACT A conducting part of a relay that acts as a switch to connect or disconnect a circuit or component.

CONTINUITY A complete path for current flow.

CURRENT The rate of flow of electrons.

CURRENT RATING The amount of current flow a device is designed to withstand.

CURRENT RELAY A relay that is operated by a predetermined amount of current flow. Current relays are often used as one type of starting relay for air-conditioning and refrigeration equipment.

DC (direct current) Current that does not reverse its direction of flow.

DEFROST CONTROL A heat pump control that reverses the flow of refrigerant in the system to heat the outside heat exchange unit and remove frost.

DELTA CONNECTION A circuit formed by connecting three electrical devices in series to form a closed loop. It is used most often in three-phase connections.

DIAC A bidirectional diode.

DIELECTRIC An electrical insulator.

DIGITAL DEVICE A device that has only two states of operation, on or off.

DIGITAL LOGIC Circuit elements connected in such a manner as to solve problems using components that have only two states of operation.

DIGITAL VOLTMETER A voltmeter that uses direct reading numerical display as opposed to a meter movement.

DIODE A two-element device that permits current to flow through it in only one direction.

DISCONNECTING MEANS (disconnect) A device or group of devices used to disconnect a circuit or device from its source of supply.

DYNAMIC BRAKING (1) Using a DC motor as a generator to produce counter torque and thereby produce a braking action. (2) Applying direct current to the stator winding of an AC induction motor to cause a magnetic braking action.

EDDY CURRENT Circular-induced current contrary to the main currents. Eddy currents are a source of heat and power loss in magnetically operated devices.

ELECTRIC CONTROLLER A device or group of devices used to govern in some predetermined manner the operation of a circuit or piece of electrical apparatus.

ELECTRICAL INTERLOCK When the contacts of one device or circuit prevent the operation of some other device or circuit.

ELECTRON One of the three major parts of an atom. The electron carries a negative charge.

ELECTRONIC CONTROL A control circuit that uses solid-state devices as control components.

EMITTER The semiconductor region of a transistor that must be connected to a polarity different from the base.

ENCLOSURE Mechanical, electrical, or environmental protection for components used in a system.

EUTECTIC ALLOY A metal with a low and sharp melting point used in thermal overload relays.

EVAPORATOR The component of an air-conditioning or refrigeration system that removes heat from the surrounding air or liquid to cause a change of state in the refrigerant.

FEEDER The circuit conductor between the service equipment, or the generator switchboard of an isolated plant, and the branch circuit overcurrent protective device.

FILTER A device used to remove the ripple produced by a rectifier.

FREQUENCY The number of complete cycles of AC voltage that occur in one second.

FULL-LOAD TORQUE The amount of torque necessary to produce the full horsepower of a motor at rated speed.

FUSE A device used to protect a circuit or electrical device from excessive current. Fuses operate by melting a metal link when current becomes excessive.

GAIN The increase in signal power produced by an amplifier.

GATE (1) A device that has multiple inputs and a single output. There are five basic types of gates, the and, or, nand, nor, and inverter. (2) One terminal of some electronic devices, such as SCRs, triacs, and field effect transistors.

HEAT ANTICIPATOR The component of a thermostat that preheats the sensing element and causes the thermostat contacts to open before the room heat has reached the set point of the thermostat.

HEAT PUMP A system that uses refrigerant to supply both heating and cooling to a dwelling.

HEAT SINK A metallic device designed to increase the surface area of an electronic component for the purpose of removing heat at a faster rate.

HERMETIC COMPRESSOR A compressor that is completely enclosed and air tight.

HERTZ (Hz) The international unit of frequency.

HOLDING CONTACTS Contacts used for the purpose of maintaining current flow to the coil of a relay.

HOLDING CURRENT The amount of current needed to keep an SCR or triac turned on.

HORSEPOWER A measure of power for electrical and mechanical devices.

HOT WIRE RELAY A type of starting relay used to disconnect the start windings of a single-phase motor.

HYSTERESIS LOOP A graphic curve that shows the value of magnetizing force for a particular type of material.

IMPEDANCE The total opposition to current flow in an electrical circuit.

INDUCED CURRENT Current produced in a conductor by the cutting action of a magnetic field.

INDUCTOR A coil.

INPUT VOLTAGE The amount of voltage connected to a device or circuit.

INSULATOR A material used to electrically isolate two conductive surfaces.

INTERLOCK A device used to prevent some action from taking place in a piece of equipment or circuit until some other action has occurred.

ISOLATION TRANSFORMER A transformer whose secondary winding is electrically isolated from its primary winding.

JUMPER A short piece of conductor used to make connection between components or a break in a circuit.

JUNCTION DIODE A diode that is made by joining together two pieces of semi-conductor material.

KICK-BACK DIODE A diode used to eliminate the voltage spike induced in a coil by the collapse of a magnetic field.

LED (light-emitting diode) A diode that will produce light when current flows through it.

LIMIT SWITCH A mechanically operated switch that detects the position or movement of an object.

LOAD CENTER Generally the service entrance. A point from which branch circuits originate.

LOCKED ROTOR CURRENT The amount of current produced when voltage is applied to a motor and the rotor is not turning.

LOCKED ROTOR TORQUE The amount of torque produced by a motor at the time of starting.

LOCKOUT A mechanical device used to prevent the operation of some other component.

LOW VOLTAGE PROTECTION A magnetic relay circuit so connected that a drop in voltage causes the motor starter to disconnect the motor from the line.

MAGNETIC CONTACTOR A contactor operated electromechanically.

MAGNETIC FIELD The space in which a magnetic force exists.

MAINTAINING CONTACT Also known as a holding or sealing contact. It is used to maintain the coil circuit in a relay control circuit. The contact is connected in parallel with the start push button.

MANUAL CONTROLLER A controller operated by hand at the location of the controller.

MICRO-FARAD A measurement of capacitance.

MICROPROCESSOR A small computer. The central processor unit is generally made from a single integrated circuit.

MODE A state or condition.

MOTOR A device used to convert electrical energy into rotating motion.

MOTOR CONTROLLER A device used to control the operation of a motor.

MULTI-SPEED MOTOR A motor that can be operated at more than one speed.

NEGATIVE One polarity of a voltage, current, or charge.

NEMA National Electrical Manufacturers Association.

NEMA RATINGS Electrical control device ratings of voltage, current, horsepower, and interrupting capability given by NEMA.

NEUTRON One of the principle parts of an atom. The neutron has no charge and is part of the nucleus.

NONINDUCTIVE LOAD An electrical load that does not have induced voltages caused by a coil. Noninductive loads are generally considered to be resistive, but can be capacitive.

NONREVERSING A device that can be operated in only one direction.

NORMALLY CLOSED The contact of a relay that is closed when the coil is de-energized.

NORMALLY OPEN The contact of a relay that is open when the coil is de-energized.

OFF-DELAY TIMER A timer that delays changing its contacts back to their normal position when the coil is de-energized.

OHMMETER A device used to measure resistance.

ON-DELAY TIMER A timer that delays changing the position of its contacts when the coil is energized.

OPERATIONAL AMPLIFIER (OP AMP) An integrated circuit used as an amplifier.

OPTOISOLATOR A device used to connect different sections of a circuit by means of a light beam.

OSCILLATOR A device used to change DC voltage into AC voltage.

OSCILLOSCOPE A voltmeter that displays a waveform of voltage in proportion to its amplitude with respect to time.

OUT-OF-PHASE The condition in which two components do not reach their positive or negative peaks at the same time.

OVERLOAD RELAY A relay used to protect a motor from damage due to overloads. The overload relay senses motor current and disconnects the motor from the line if the current is excessive for a certain length of time.

PANELBOARD A metallic or nonmetallic panel used to mount electrical controls, equipment, or devices.

PARALLEL CIRCUIT A circuit that contains more than one path for current flow.

PEAK-INVERSE/PEAK-REVERSE VOLTAGE The rating of a semi-conductor device that indicates the maximum amount of voltage in the reverse direction that can be applied to the device.

PEAK-TO-PEAK VOLTAGE The amplitude of AC voltage measured from its positive peak to its negative peak.

PEAK VOLTAGE The amplitude of voltage measured from zero to its highest value.

PERMANENT SPLIT-CAPACITOR MOTOR A single-phase induction motor similar to the capacitor start motor except that the start windings and the starting capacitor remains connected in the circuit during normal operation.

PHASE SHIFT A change in the phase relationship between two quantities of voltage or current.

PILOT DEVICE A control component designed to control small amounts of current. Pilot devices are used to control larger control components.

PNEUMATIC TIMER A device that uses the

displacement of air in a bellows or diaphragm to produce a time delay.

POLARITY   The characteristic of a device that exhibits opposite quantities within itself: positive and negative.

POTENTIOMETER   A variable resistor with a sliding contact that is used as a voltage divider.

POWER FACTOR   A comparison of the true power (watts) to the apparent power (volt amps) in an AC circuit.

POWER RATING   The rating of a device that indicates the amount of current flow and voltage drop that can be permitted.

PRESSURE SWITCH   A device that senses the presence or absence of pressure and causes a set of contacts to open or close.

PRINTED CIRCUIT   A board on which a pre-determined pattern of printed connections has been made.

PROTON   One of the three major parts of an atom. The proton has a positive charge.

PUSH BUTTON   A pilot control device operated manually by being pushed or pressed.

REACTANCE   The opposition to current flow in an AC circuit offered by pure inductance or pure capacitance.

RECTIFIER   A device or circuit used to change AC voltage into DC voltage.

REGULATOR   A device that maintains a quantity at a predetermined level.

RELAY   A magnetically-operated switch that may have one or more sets of contacts.

REMOTE CONTROL   Controls the functions of some electrical device from a distant location.

RESISTANCE   The opposition to current flow in an AC or DC circuit.

RESISTANCE START INDUCTION RUN MOTOR   One type of split-phase motor that uses the resistance of the start winding to produce a phase shift between the current in the start winding and the current in the run winding.

RESISTOR   A device used to introduce some amount of resistance into an electrical circuit.

RHEOSTAT   A variable resistor.

RMS VALUE   The value of AC voltage that will produce as much power when connected across a resistor as a like amount of DC voltage.

SATURATION   The maximum amount of mag-

netic flux a material can hold.

SCHEMATIC   An electrical drawing showing components in their electrical sequence without regard for physical location.

SCR   (SILICON-CONTROLLED RECTI-FIER)   A semi-conductor device tht can be used to change AC voltage into DC voltage. The gate of the SCR must be triggered before the device will conduct current.

SEMI-CONDUCTOR   A material that contains four valence electrons and is used in the production of solid-state devices.

SENSING DEVICE   A pilot device that detects a quantity and converts it into an electrical signal.

SERIES CIRCUIT   A circuit that contains only one path for current flow.

SERVICE   The conductors and equipment necessary to deliver energy from the electrical supply system to the premises served.

SERVICE FACTOR   An allowable overload for a motor indicated by a multiplier that, when applied to a normal horsepower rating, indicates the permissible loading.

SHADED-POLE MOTOR   An AC induction motor that develops a rotating magnetic field by shading part of the stator windings with a shading loop.

SHADING LOOP   A large copper wire or band connected around part of a magnetic pole piece to oppose a change of magnetic flux.

SHORT CIRCUIT   An electrical circuit that contains no resistance to limit the flow of current.

SHORT CYCLING   The starting and stopping of a compressor in rapid succession.

SINE-WAVE VOLTAGE   A voltage waveform whose value at any point is proportional to the trigonometric sine of the angle of the generator producing it.

SLIP   The difference in speed between the rotating magnetic field and the speed of the rotor in an induction motor.

SNAP-ACTION   The quick opening and closing action of a spring loaded contact.

SOLENOID   A magnetic device used to convert electrical energy into linear motion.

SOLENOID VALVE   A valve operated by an electric solenoid.

SOLID-STATE DEVICE   An electronic component constructed from semi-conductor material.

SPLIT-PHASE MOTOR A type of single-phase motor that uses resistance or capacitance to cause a shift in the phase of the current in the run winding and the current in the start winding. The three primary types of split-phase motors are resistance start induction run, capacitor start induction run, and permanent split-capacitor motor.

STAGING THERMOSTAT A thermostat that contains more than one set of contacts that operate at different times in accord with the temperature.

STARTER A relay used to connect a motor to the power line.

STATOR The stationary winding of an AC motor.

STEP-DOWN TRANSFORMER A transformer that produces a lower voltage at its secondary than is applied to its primary.

STEP-UP TRANSFORMER A transformer that produces a higher voltage at its secondary than is applied to its primary.

SURGE A transient variation in the current or voltage at a point in the circuit. Surges are generally unwanted and temporary.

SWITCH A mechanical device used to connect or disconnect a component or circuit.

SYNCHRONOUS SPEED The speed of the rotating magnetic field of an AC induction motor.

TEMPERATURE RELAY A relay that functions at a predetermined temperature. Generally used to protect some other component from excessive temperature.

TERMINAL A fitting attached to a device for the purpose of connecting wires to it.

THERMISTOR A resistor that changes its resistance with a change of temperature.

THYRISTOR An electronic component that has only two states of operation: on or off.

TORQUE The turning force developed by a motor.

TRANSDUCER A device that converts one type of energy into another type of energy. Example: A solar cell converts light into electricity.

TRANSFORMER An electrical device that changes one value of AC voltage into another value of AC voltage.

TRANSISTOR A solid-state device made by combining three layers of semi-conductor material

together. A small amount of current flow through the base-emitter can control a larger amount of current flow through the collector-emitter.

TRIAC A bidirectional thyristor used to control AC voltage.

TROUBLESHOOT To locate and eliminate problems in a circuit.

VALENCE ELECTRONS Electrons located in the outer orbit of an atom.

VARIABLE RESISTOR A resistor whose resistance value can be varied between its minimum and maximum values.

VARISTOR A resistor that changes its resistance value with a change of voltage.

VOLTAGE An electrical measurement of potential difference, electrical pressure, or electromotive-force (EMF).

VOLTAGE DROP The amount of voltage required to cause an amount of current to flow through a certain resistance.

VOLTAGE RATING A rating that indicates the amount of voltage that can be safely connected to a device.

VOLTAGE REGULATOR A device or circuit that maintains a constant value of voltage.

VOLTMETER An instrument used to measure a level of voltage.

VOLT-OHM-MILLIAMMETER (VOM) A test instrument so designed that it can be used to measure voltage, resistance, or milliamperes.

WATT A measure of true power.

WAVEFORM The shape of a wave obtained by plotting a graph of voltage with respect to time.

WIRING DIAGRAM An electrical diagram used to show components in their approximated physical location with connecting wires.

WYE CONNECTION A connection of three components made in such a manner that one end of each component is connected. This connection is generally used to connect devices to a three-phase power system.

ZENER DIODE A diode that has a constant voltage drop when operated in the reverse direction. Zener diodes are commonly used as voltage regulators in electronic circuits.

ZONE Generally refers to heating or cooling a certain section of a dwelling.

INDEX

Note: Numbers followed by a *t* indicate tabular material.

Voltage, control of, triac for, 307
current and, 45, 51–52
defined, 7–9
drop, 37–42
drop in, testing for, 42
rating, 40
transformers and, 116–18
Voltmeter, interpreting, 15–16
types of, 12–15

W

Wattage, defined, 9–10
Wire, cross-sectional area of, 37
gauges of, 39*t*
resistance, 37
resistance of, 37–38
size, 37–42
sizing of, 40–41
Wiring diagrams, 190–96
development of, 198–203
reading of, 193–96
Wye connection, 28–29

Z

Zero switching, 113–14